LEARNING TARGETS
for Numeracy

Number

Key Stage 2

David Clemson

Wendy Clemson

Derek Kassem

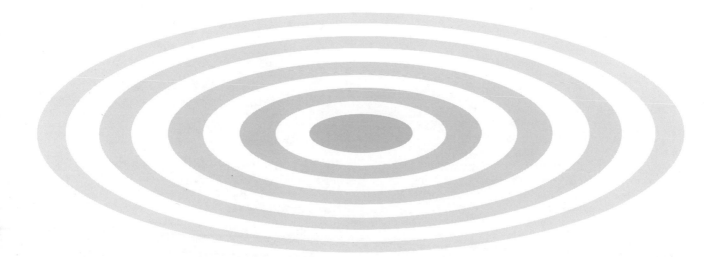

Stanley Thornes (Publishers) Ltd

Stanley Thornes for TEACHERS:
BLUEPRINTS • PRIMARY COLOURS • LEARNING TARGETS

Stanley Thornes for Teachers publishes practical teacher's ideas books and photocopiable resources for use in primary schools. Our three key series, **Blueprints**, **Primary Colours** and **Learning Targets** together provide busy teachers with unbeatable curriculum coverage, inspiration and value for money. We mail teachers and schools about our books regularly. To join the mailing list simply photocopy and complete the form below and return using the **FREEPOST** address to receive regular updates on our new and existing titles. You may also like to add the name of a friend who would be interested in being on the mailing list. Books can be bought by credit card over the telephone and information obtained on (01242) 267280.

Please add my name to the *Stanley Thornes for* TEACHERS mailing list.

Mr/Mrs/Miss/Ms _____

Address _____

_____ postcode _____

School address _____

_____ postcode _____

Please also send information about *Stanley Thornes for* TEACHERS to:

Mr/Mrs/Miss/Ms _____

Address _____

_____ postcode _____

To: Marketing Services Dept., Stanley Thornes Ltd, FREEPOST (GR 782), Cheltenham, GL50 1BR

Text © David Clemson, Wendy Clemson, Derek Kassem 1998.

The right of David Clemson, Wendy Clemson and Derek Kassem to be identified as authors of this work has been asserted by them in accordance with the Copyright, Designs and Patents Act 1988.

First published in 1998 by
Stanley Thornes Publishers Ltd
Ellenborough House
Wellington Street
Cheltenham GL50 1YW

00 01 02 / 10 9 8 7 6 5 4

A catalogue record for this book is available from the British Library.

ISBN 0–7487–3597–6

Typeset by Tech-Set, Gateshead, Tyne & Wear
Illustrations by Andrew Keylock
Printed and bound in Great Britain by Redwood Books, Trowbridge, Wiltshire

CONTENTS

Welcome to
LEARNING TARGETS

Learning Targets is a series of practical teacher's resource books written to help you to plan and deliver well-structured, professional lessons in line with all the relevant curriculum documents.

Each Learning Target book provides exceptionally clear lesson plans which cover the whole of its stated curriculum plus a large bank of carefully structured copymasters. Links to the key curriculum documents are provided throughout to enable you to plan effectively.

The Learning Targets series has been written in response to the challenge confronting teachers not just to come up with teaching ideas which cover the curriculum, but to ensure that they deliver high quality lessons every lesson, with the emphasis on raising standards of pupil achievement.

The recent thinking from OFSTED, and the National Literacy and Numeracy Strategies on the key factors in effective teaching has been built into the structure of Learning Targets. These might briefly be summarised as follows:

➽ that effective teaching is active teaching directed to very clear objectives
➽ that good lessons are delivered with pace, rigour and purpose
➽ that good teaching requires a range of strategies – including interactive whole class sessions
➽ that ongoing formative assessment is essential to plan children's learning
➽ that differentiation is necessary but that it must be realistic.

The emphasis in Learning Targets is on absolute clarity. We have written and designed the books to enable you to access and deliver effective lessons as easily as possible, with the following aims:

➽ to plan and deliver rigorous, well-structured lessons
➽ to set explicit targets for achievement in every lesson that you teach
➽ to make the children aware of what they are going to learn
➽ to put the emphasis on direct, active teaching every time
➽ to make effective use of time and resources
➽ to employ the full range of recommended strategies whole-class, group and individual work
➽ to differentiate for ability groups realistically
➽ to use ongoing formative assessment to plan your next step
➽ to have ready access to usable pupil copymasters to support your teaching.

The page opposite provides an at-a-glance guide to the key features of the Learning Targets lessons and explains how they will enable you deliver effective lessons. The key to symbols on the lesson plans is set out here. ➽➤

How to deliver structured lessons with pace, rigour and purpose

Explicit targets for achievement in every lesson

The concise subject knowledge you need

Crystal clear lesson plan layouts

The full range of teaching strategies

Rigorous and practical activities

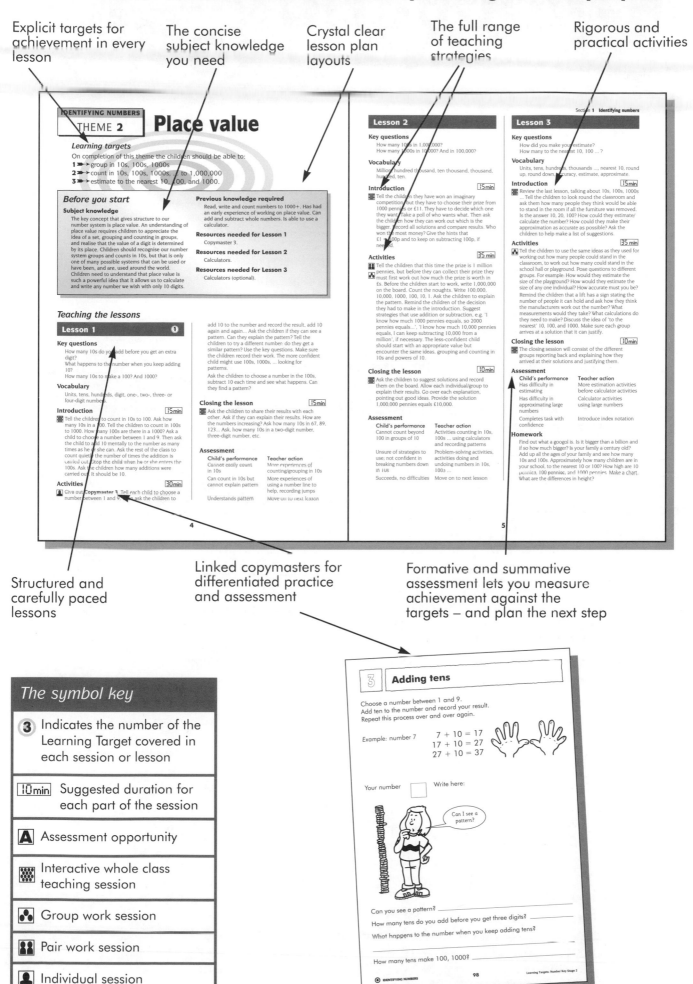

Structured and carefully paced lessons

Linked copymasters for differentiated practice and assessment

Formative and summative assessment lets you measure achievement against the targets – and plan the next step

The symbol key

3	Indicates the number of the Learning Target covered in each session or lesson
10min	Suggested duration for each part of the session
A	Assessment opportunity
	Interactive whole class teaching session
	Group work session
	Pair work session
	Individual session

v

INTRODUCTION

Learning Targets: *Number Key Stage* 2 includes lessons on all of the main ideas in number for children aged 7–11 (Years 3–6/P4–7). Together with its companion book, *Learning Targets*: *Shape, Space and Measures Key Stage* 2, it offers support for the teaching of all the key features of mathematics suitable for children of this age group. In planning and writing this book the authors have not only sought to meet the requirements of the National Curriculum, and Curriculum and Assessment in Scotland: National Guidelines: Mathematics 5–14, but have also borne in mind the fact that there are currently demands for teachers to use direct and whole class teaching as a regular part of their teaching repertoire, and that teachers need to be aware of and address the imperatives highlighted in the National Numeracy Project.

This book and its companion volume do not, of course, constitute a complete scheme. They cannot provide you with all the resources needed for every mathematics session. As it covers all of the main ideas in number work, however, this book is a backbone resource for mathematics teaching. There are some lessons at each level of work appropriate for the Key Stage. This text can therefore be seen as an extremely valuable and effective aid to the delivery of directly taught lessons in number. It contains a series of well-structured, detailed and specific lesson plans, backed by linked copymasters, which you can use to teach lessons in line with National Curricula and the National Numeracy Project.

As each of the four mathematics books in the series addresses work at a whole Key Stage (either Years R–2/P1–3 or Years 3–6/P4–7), it is necessary to select lessons at the appropriate level. To help you to do this, the books are organised into sections which each contain a number of themes. There is progression from the start of each section to its end. Within each theme there are three lessons which also offer a progression, and the lessons should therefore be taught in order.

How this book is organised

Sections

This book is organised into six sections; these being: Identifying Numbers; Addition and Subtraction; Fractions; Multiplication and Division; Mental Arithmetic and Number Patterns; and Handling Data. This last section concentrates on data collection and analysis in the context of number work.

At the start of each section you will find a short overview of the mathematics ideas which we see as important in offering the children appropriate learning opportunities. A section is divided into a number of themes, each with its own set of clear learning targets. There is a progression within each section. To conclude each section there is a set of extension ideas. These can be used in any mathematics session where the key ideas related to this section are being worked on. They may be incorporated within a programme of lessons, used in sessions that immediately follow learning target lessons, used in sessions designated as mathematical investigation sessions or used as additional homework activities.

Themes

The order of the themes within each section has been arranged to offer progression. Thus, in Section 1, Identifying Numbers, there are four themes. The first theme, 'Using the number line', may be seen as more appropriate for children at an earlier stage in their mathematical education than the next theme, 'Place value'. Thus 'Using the number line' might be the theme tackled in Year 3/P4, while the theme on place value may form part of the course for a Year 4/P5 class of children. The themes which follow 'Using the number line' and 'Place value' place a increasing demand on children's mathematical skills and knowledge, and might therefore be taught to Upper Juniors (in Year 5–6/P6–7).

Within each theme the lessons are also sequenced to provide more demand as the children move on from the first, to second and then third. The learning targets state explicitly what the children should know or be able to do by the end of each lesson. The learning targets provide you with a clear set of assessable objectives.

The themes in a section together form an overall set of lesson plans for a mathematics topic. The themes are free-standing. It is also possible for you to choose lessons from within a theme as free-standing lessons. At the end of each lesson there are descriptions of children's performances and suggested teacher actions. At the end of each theme there are suggestions for homework activities.

The lesson plans with each theme are very specific and detailed in their teaching suggestions, written to allow you to undertake direct teaching to clear objectives. Some lessons have accompanying copymasters which are completely integrated into the teaching activities.

National Curricula and Numeracy

The lessons in this book have been written to meet the time demands of the 'numeracy hour'. The mathematical ideas match the required range of work in National Curricula and those seen as important in the National Numeracy project. There are charts showing how themes in the book fit the programmes of study in the National Curriculum (England and Wales) and areas in the National Numeracy Project recommendations on pages viii–ix.

The need to revisit mathematics topics as children progress through Key Stage 2 (from Year 3–6/P4–7) has meant that the book is organised into sections which, as we have already indicated, can be used flexibly across the whole age range. We have written the lessons so that the teacher can differentiate between children's learning by the outcomes of their work.

Each theme can provide the material for a string of numeracy hours. Every teacher will interpret the demands of the numeracy hour in the light of their own situation and the structure of the book allows for this. To plan your number work we suggest that you consult the appropriate section and theme title to locate the lesson you want when you wish to offer a direct teaching session to your class.

The learning targets for each theme have been mapped against the Programmes of Study in the National Curriculum for England and Wales, Levels 3–5 and areas in the National Numeracy Project Recommendations. This chart is presented on pages viii–ix.

The learning targets for each theme have also been mapped against the statements in the attainment targets in Curriculum and Assessment in Scotland: National Guidelines: Mathematics 5–14 at Levels B, C and D. This chart is presented on page x. Teachers in Scotland can therefore be confident that the lessons in this book meet the requirements to which they are working.

Theme No.	Place value/ order/rounding	Number properties	Fractions/decimals/ percentages/ratio	Understand + and −	Recall + and − facts	Mental strategies + and −	Pencil and paper + and −	× and ÷	Recall × and ÷ facts	Mental strategies × and ÷	Pencil and paper × and ÷	Checking	Making decisions	Reasoning about numbers	Problems: real life/ money	Data handling/ probability
1	●	●	●	●	●	●	●									
2	●															
3	●	●	●													
4	●	●				●							●	●		
5				●	●	●	●									
6				●	●	●	●									
7	●			●	●	●	●									
8				●	●	●	●									
9				●		●	●									
10				●		●	●									
11			●	●		●	●									
12															●	
13	●		●	●		●	●							●		
14			●													
15			●													
16			●													
17			●													
18								●	●	●	●				●	
19								●	●	●	●	●	●	●	●	
20	●							●	●	●	●					
21								●		●	●				●	
22	●							●			●				●	
23	●							●			●					
24			●					●			●				●	
25			●					●			●				●	
26			●												●	
27			●												●	
28				●	●	●										
29				●	●	●										
30				●						●			●	●	●	
31			●												●	
32								●	●	●	●	●			●	
33				●		●				●						
34																●
35																●
36																●
37																●
38																●
39																●
40																●
41																●

Curriculum planners
Programmes of study in the National Curriculum (England and Wales)

Number

Theme No.	2a	2b	2c	3a	3b	3c	3d	3e	3f	3g	3h	4a	4b	4c
1	●	●	●	●		●		●						
2	●										●			
3	●	●		●										
4	●			●									●	
5						●								
6						●								
7	●					●	●							
8						●								
9							●							
10							●							
11		●								●				
12						●								
13	●	●				●								
14			●											
15			●											
16			●											
17			●											
18						●	●	●				●		
19				●		●	●	●	●	●		●	●	●
20						●	●	●			●			
21							●					●		
22	●						●					●		
23	●					●	●	●			●			
24		●								●	●	●		
25		●								●	●	●		
26			●									●		
27			●									●		
28						●	●							
29				●		●	●							
30				●		●								
31		●										●		
32						●	●					●		
33				●		●								
34														
35														
36														
37														
38														
39						●						●		
40														
41				●										

Handling Data

Theme No	2a	2b	2c	2d	3a	3b	3c
1							
2							
3							
4							
5							
6							
7							
8							
9							
10							
11							
12							
13							
14							
15							
16							
17							
18							
19							
20							
21							
22							
23							
24							
25							
26							
27							
28							
29							
30							
31							
32							
33							
34	●	●					
35	●	●					
36	●	●		●			
37			●				
38					●		
39	●	●		●			
40			●				
41					●	●	●

Curriculum planners
Scottish guidelines planner

LEVEL B	LEVEL C	LEVEL D
INFORMATION HANDLING ATTAINMENT TARGET		
Collect		
Conduct a class survey Theme 35	Obtain information (variety of sources) Themes 4, 35, 38	Select information sources Themes 4, 40, 41
Organise		
Use a tally sheet Theme 34	Enter data in a table Themes 35, 38	Use diagrams and tables Themes 40, 41
Display		
Construct a bar graph Theme 34	Construct table or chart Theme 36 Construct a bar graph with multiple units Theme 36	Construct graphs Theme 39
Interpret		
	From displays and databases Themes 4, 36, 37, 38	From a range of displays and databases Themes 4, 37, 38, 40, 41
NUMBER, MONEY AND MEASUREMENT* ATTAINMENT TARGET		
Range and type of numbers		
Whole numbers up to 100 then 1000 Themes 2, 3 Quarters Themes 14, 15	Whole numbers up to 10 000 Themes 2, 3, 4, 13, Thirds, fifths, eighths, tenths and simple equivalences Theme 16 Decimals to two places Themes 11, 17	Whole numbers to 100 000, a million Themes 4, 13 Fractions Theme 17 Percentages, decimals to 2 places Theme 7
Money		
Use coins to £1 Themes 12, 31	Use coins/notes to £25 Themes 12, 31	
Add and subtract		
Mentally 0–20 Themes 1, 5, 6, 28, 29 2 digits (without calculator) Themes 1, 3, 7, 8, 29, 29 2 digit numbers added to or subtracted from 3 digit numbers (with calculator) Themes 3, 9, 10	Mentally one digit to or from numbers up to 3 digits Theme 29 Subtraction by 'adding on' Theme 29 Without calculator, whole numbers with 2 digits added to or subtracted from 3 digits Themes 3, 33 With calculator, for 3 digit whole numbers Money applications up to £20 Theme 12	Mentally, 2 digit whole numbers Theme 29 Without calculator for 4 digits with 2 decimal places (easy) Theme 33
Multiply and divide		
Mentally 2, 3, 4, 5, 10 Theme 1 Without calculator, 2 digits multiplied by 2, 3, 4, 5, 10 Themes 19, 20, 21 With calculator for 2 digit numbers multiplied/divided by any digit Themes 19, 20, 21 Money applications up to £1 Theme 31	Mentally all tables to 10 Theme 18 Mentally 2 or 3 digit numbers by 10 Themes 22, 23 Without calculator, 2 digit numbers by 1 digit Themes 9, 20, 21, 22, 23 With calculator 2 or 3 digits by 1 or 2 digit numbers by 1 or 2 digits Themes 19, 20, 21, 22, 23 Money applications up to £20 Theme 31	Mentally whole numbers by single digits (easy) Theme 18 Mentally for 4 digit numbers including decimals by 10 or 100 Themes 22, 23, 24 Without calculator for 4 digits with up to 2 decimal places by a single digit Themes 22, 23, 24, 25 With calculator for 4 digits with up to 2 decimal places, by 2 digits Themes 24, 25 Applications in money
Fractions, percentages and ratios		
Find halves and quarters Themes 14, 15	Find simple fractions Themes 15, 16	Work with simple fractions Themes 17, 26, 27
Patterns and sequences		
Whole numbers sequences within 100 Theme 1	Patterns within multiplication tables Themes 1, 32, 33	More complex sequences Themes 32, 33
Functions and equations		
Find missing numbers Theme 30	Use simple 'function machines' for doubling, halving, adding and subtraction Theme 30	Recognise and explain simple relationships between two sets of numbers or objects Theme 39

*For coverage of measurement see the companion book: *Learning Targets For Numeracy: Shape, space and measures Key Stage 2.*

IDENTIFYING NUMBERS

There are three key ingredients for children to handle numbers and appreciate their properties successfully. They need to understand the language of numbers in relation to the symbols and conventions we have for naming and manipulating number. Place value is the central attribute of our number system and those who do not gain good insights into this concept and how place value works will have continuing problems. The children need to appreciate that, especially in the primary years, our numbers fit on the number line and this line, even when held visually in memory, is a marvellous way of understanding operations with and on numbers.

The history of mathematical ideas relating to number is worth studying with children. Not only does this give a human feel to mathematics but it also, in some respects, mirrors the child's own growing sophistication with numbers. To explore number names in English and to contrast this with other languages gives a real sense of why we have special words to denote, in our case, 'ten'. So 'seventeen' can be seen as 'seven and ten'. In just the same way the French talk about *dix-sept* which we could read as 'ten and seven'.

However, differences are important too. Where we have started our use of 'teen' after 12, the French start at 17. Of course the decimal (from the Latin) system is only one base we can work in. In Britain we had, until comparatively recently, systems for money using 12s and 20s, and for volume 8s, with gallons and pints – and we still use pints. It is also the case that other societies have used other bases – 2 and 5 being common; and the Babylonians used 60 as their key number.

This use of different bases links well to the development of an appreciation of place value. The system we have of just ten symbols for numbers is so simple and versatile that it is interesting to see just how long humanity took to achieve a genuine place value system. The key feature is, of course, the invention of 'zero' as a place holder. With that development, we could abandon the creation of more and more symbols to represent special numbers.

It is vital, therefore, that young children get a continuing and practically-based grounding in the way in which our place value system works. Stress should be placed on the function of zero for here there are problems of language when people talk about '0' as nothing, nil, nought, and 'O'.

Finally, the versatility of the number line means that we can construct number lines with any dimensions we wish; so it is not used just for learning simple number bonds but also supports an appreciation of fractions and decimals. The number line has so many physical forms in everyday life to do with measuring, from tapes through rulers to micrometers. Even where we cannot locate a number on the number line we can still use it to obtain a good approximation. Such irrational numbers as $\sqrt{2}$, whilst having no solution, can be located in a region on the number line.

Using the number line

Learning targets

On completion of this theme the children should be able to:

1 ➤➤ use a number line in support of addition and subtraction of natural numbers
2 ➤➤ use a number line in support of their learning of multiplication tables
3 ➤➤ carry out an investigation using a number line.

Before you start

Subject knowledge

This theme contains lesson ideas that can be carried through sequentially or can be selected and used in support of other areas of number work: the choice is yours. The number line is one of the big ideas in mathematics. It holds within it so much of our number system and appreciation of number. Ideas such as infinity, relationships, patterns, and sequences can all be generated out of using and thinking about a number line. Number lines can and should be used at all levels of the mathematics work of primary age children. However, there are a couple of imperatives when it comes to their construction and use.

The children need, from the start, to see that a number line is about continuous not discrete information. So to draw or otherwise symbolise a number line by labelling the spaces between points is wrong. If this happens then it is so much harder to persuade the children that the space between, say, 7 and 8, has an infinite number of points which can be labelled – and the line between those labels can be further explored. So it is the points on the number line we should label, not the spaces between them.

Previous knowledge required

Number names and symbols, one-to-one correspondence, conservation of number, components of numbers, addition and subtraction of whole numbers, multiplication tables.

Resources needed for Lesson 1

Number lines made up from Copymaster 62 (one for each child), Copymaster 1, coloured counters (several counters of each of several colours).

Resources needed for Lesson 2

Copymaster 2, coloured counters (several counters of each of several colours).

Resources needed for Lesson 3

Copymaster 60.

Teaching the lessons

Lesson 1 ①

Key questions

What sorts of numbers are these?
Can you see a pattern?
What lies between?

Vocabulary

Number, number line, addition, subtraction, multiplication, odd and even.

Introduction 15 min

▓ Use this lesson to revise what the children should already know, or have encountered before, about the properties of numbers. Start by asking the children to tell you as much as they can about a number, e.g. 7. List their ideas on the board so that you can return to them at the end of the lesson.

Activities 25 min

👤 Give out **Copymaster 1** and the number lines. Go around the class as the children develop their responses to the questions. If they 'just know' the correct answer and do not need to use the number line tell them that is fine.

Closing the lesson 10 min

▓ Revisit the number 7 list and see if there are things the children would like to add. Get them to tell you and the rest of the class any interesting things they noticed or discovered whilst doing Copymaster 1.

Assessment

Child's performance	Teacher action
Insecure in some aspects of number proportion	Check which aspects caused most difficulty then offer some focused opportunities in those areas
Satisfactory response	Move on to next lesson
Ready facility with all of the ideas	Move on to next lesson

Lesson 2 ②

Key questions

What numbers do we get?
Can you see a pattern?
Can you see any common patterns?

Vocabulary

Number line, multiplication, pattern, series, doubling and halving.

Introduction [10 min]

Remind the children of the work they have done on number lines, and on multiplication tables – a quick tables warm up might be appropriate. Tell them that in this lesson they are going to use a number line to support their knowledge of some tables. Using a 0–20 number line on the board go through the 2× table and go along the line counting on in 2s. Use language like, '2 lots of 3', '3 lots of 2' and similar examples.

Activities [25 min]

Give out **Copymaster 2**. The emphasis here is on the children taking responsibility for developing the ideas. They have to use their knowledge of times tables to see what patterns they can find in the tables, and to indicate these on the number line. Encourage their thinking by using the 5× and 10× tables to illustrate what you want them to do. If some do this quickly then you might encourage them to look at patterns and sequences using division.

Closing the lesson [15 min]

Get ideas from each pair and get them to explain their ideas to the others if they are not immediately clear. Write a selection of the ideas on the board as they emerge and use these as an opportunity to recap on points at the end.

Assessment

Child's performance	Teacher action
Relatively few different ideas	Give more time/practice following what they have heard from their peers in the closing part of the lesson
Good range of ideas	Move on to next lesson
Good range of ideas and develops some in respect of division	Move on to next lesson

Lesson 3 ③

Key questions

What results from the addition/subtraction of these consecutive numbers?
How can you make this using consecutive numbers?
In how many ways can you make ... using pairs of numbers?

Vocabulary

Addition, subtraction, number line.

Introduction [15 min]

The children should now be very familiar with using number lines. Draw a 0–20 number line on the board emphasising that the points are labelled not the spaces. Choose some consecutive pairs of numbers and ask the children what they sum to. Ask what the difference between these consecutive pairs is. Get the children to make a 0–20 number line using **Copymaster 60**.

Activities [20 min]

Working mentally, the children should list the results of adding all of the consecutive pairs of numbers on the number line. You might choose to do this with the class before moving to pair up the children. They then have to find all of the pairs of numbers on the number line that will make 8 (or 6 or 9, or whatever target number you choose). How many pairs can they find? Some children might do more than one target number. Some might use points on the number line that are not whole numbers.

Closing the lesson [10 min]

Get rapid mental responses to the addition of consecutive numbers. Ask for suggestions as to the different ways to make 8 (or your chosen target number). Write them on the board. Encourage the children to see that using pairs to make target numbers is a good way of practising their number bond work.

Assessment

Child's performance	Teacher action
Copes with consecutive pairs but not with target number/s	Revise early work on common addition and subtraction number bonds
Completes tasks satisfactorily but misses some combinations	Give more practice on finding target numbers
Completes tasks satisfactorily	The learning targets for this theme have been met

Homework

Using a 0–20, 0–50, or 0–100 number line the children should put coloured dots next to points that are products in particular multiplication tables. Ask them to find all the numbers from 0–20 that can be made by adding consecutive numbers.

Place value

Learning targets

On completion of this theme the children should be able to:

1 ➡ group in 10s, 100s, 1000s
2 ➡ count in 10s, 100s, 1000s, ... to 1,000,000
3 ➡ estimate to the nearest 10, 100, and 1000.

Before you start

Subject knowledge

The key concept that gives structure to our number system is place value. An understanding of place value requires children to appreciate the idea of a set, grouping and counting in groups, and realise that the value of a digit is determined by its place. Children should recognise our number system groups and counts in 10s, but that is only one of many possible systems that can be used or have been, and are, used around the world. Children need to understand that place value is such a powerful idea that it allows us to calculate and write any number we wish with only 10 digits.

Previous knowledge required

Read, write and count numbers to 1000+. Has had an early experience of working on place value. Can add and subtract whole numbers. Is able to use a calculator.

Resources needed for Lesson 1

Copymaster 3.

Resources needed for Lesson 2

Calculators.

Resources needed for Lesson 3

Calculators (optional).

Teaching the lessons

Lesson 1 ①

Key questions

How many 10s do you add before you get an extra digit?
What happens to the number when you keep adding 10?
How many 10s to make a 100? And 1000?

Vocabulary

Units, tens, hundreds, digit, one-, two-, three- or four-digit numbers.

Introduction 15 min

▨ Tell the children to count in 10s to 100. Ask how many 10s in a 100. Tell the children to count in 100s to 1000. How many 100s are there in a 1000? Ask a child to choose a number between 1 and 9. Then ask the child to add 10 mentally to the number as many times as he or she can. Ask the rest of the class to count quietly the number of times the addition is carried out. Stop the child when he or she enters the 100s. Ask the children how many additions were carried out. It should be 10.

Activities 30 min

▨ Give out **Copymaster 3**. Tell each child to choose a number between 1 and 9. Then ask the children to

add 10 to the number and record the result, add 10 again and again... Ask the children if they can see a pattern. Can they explain the pattern? Tell the children to try a different number: do they get a similar pattern? Use the key questions. Make sure the children record their work. The more confident child might use 100s, 1000s, ... looking for patterns.

Ask the children to choose a number in the 100s, subtract 10 each time and see what happens. Can they find a pattern?

Closing the lesson 15 min

▨ Ask the children to share their results with each other. Ask if they can explain their results. How are the numbers increasing? Ask how many 10s in 67, 89, 123... Ask, how many 10s in a two-digit number, three-digit number, etc.

Assessment

Child's performance	Teacher action
Cannot easily count in 10s	More experiences of counting/grouping in 10s
Can count in 10s but cannot explain pattern	More experiences of using a number line to help, recording jumps
Understands pattern	Move on to next lesson

Lesson 2

Key questions
How many 100s in 1,000,000?
How many 1000s in 10,000? And in 100,000?

Vocabulary
Million, hundred thousand, ten thousand, thousand, hundred, ten.

Introduction `15 min`
▦ Tell the children they have won an imaginary competition, but they have to choose their prize from 1000 pennies or £11. They have to decide which one they want. Take a poll of who wants what. Then ask the children how they can work out which is the bigger. Record all solutions and compare results. Who won the most money? Give the hints that £1 = 100p and to keep on subtracting 100p, if needed.

Activities `35 min`
▦▦ Tell the children that this time the prize is 1 million pennies, but before they can collect their prize they must first work out how much the prize is worth in £s. Before the children start to work, write 1,000,000 on the board. Count the noughts. Write 100,000, 10,000, 1000, 100, 10, 1. Ask the children to explain the pattern. Remind the children of the decision they had to make in the introduction. Suggest strategies that use addition or subtraction, e.g. 'I know how much 1000 pennies equals, so 2000 pennies equals...', 'I know how much 10,000 pennies equals, I can keep subtracting 10,000 from a million', if necessary. The less-confident child should start with an appropriate value but encounter the same ideas, grouping and counting in 10s and powers of 10.

Closing the lesson `10 min`
▦ Ask the children to suggest solutions and record them on the board. Allow each individual/group to explain their results. Go over each explanation, pointing out good ideas. Provide the solution 1,000,000 pennies equals £10,000.

Assessment

Child's performance	Teacher action
Cannot count beyond 100 in groups of 10	Activities counting in 10s, 100s ... using calculators and recording patterns
Unsure of strategies to use; not confident in breaking numbers down in 10s	Problem-solving activities, activities doing and undoing numbers in 10s, 100s ...
Succeeds, no difficulties	Move on to next lesson

Lesson 3

Key questions
How did you make your estimate?
How many to the nearest 10, 100 ... ?

Vocabulary
Units, tens, hundreds, thousands ..., nearest 10, round up, round down, accuracy, estimate, approximate.

Introduction `15 min`
▦ Review the last lesson, talking about 10s, 100s, 1000s ... Tell the children to look round the classroom and ask them how many people they think would be able to stand in the room if all the furniture was removed. Is the answer 10, 20, 100? How could they estimate/calculate the number? How could they make their approximation as accurate as possible? Ask the children to help make a list of suggestions.

Activities `35 min`
▦ Tell the children to use the same ideas as they used for working out how many people could stand in the classroom, to work out how many could stand in the school hall or playground. Pose questions to different groups. For example: How would they estimate the size of the playground? How would they estimate the size of any one individual? How accurate must you be?

Remind the children that a lift has a sign stating the number of people it can hold and ask how they think the manufacturers work out the number? What measurements would they take? What calculations do they need to make? Discuss the idea of 'to the nearest' 10, 100, and 1000. Make sure each group arrives at a solution that it can justify.

Closing the lesson `10 min`
▦ The closing session will consist of the different groups reporting back and explaining how they arrived at their solutions and justifying them.

Assessment

Child's performance	Teacher action
Has difficulty in estimating	More estimation activities before calculator activities
Has difficulty in approximating large numbers	Calculator activities using large numbers
Completes task with confidence	Introduce index notation

Homework
Find out what a googol is. Is it bigger than a billion and if so how much bigger? Is your family a century old? Add up all the ages of your family and see how many 10s and 100s. Approximately how many children are in your school, to the nearest 10 or 100? How high are 10 pennies, 100 pennies, and 1000 pennies. Make a chart. What are the differences in height?

THEME 3 | Directed numbers

Learning targets

On completion of this theme the children should be able to:
1 ➡ recognise directed numbers
2 ➡ use directed numbers in a real context
3 ➡ add and subtract directed numbers on a number line.

Before you start

Subject knowledge

Most people think of the cardinal aspect of number, e.g. 7 = the set of 7 objects. It is very difficult to imagine a set of 7 negative items; thus it is better to think of directed numbers as part of the number line. The most common example of directed numbers usually given is bank balances. They are always moving up and down (into the red (negative), and into the black (positive)) and can all be represented on a number line. Using the idea of bank balances, if we are £10 in the red, that is have a balance of −£10, to get to zero we must add £10, −10 + 10 = 0. Or if we put £15 in the bank to rectify the problem then −10 + (+15) = +5. Equally, if we had £10 in the bank and withdrew £6, 10 +(−6) = +4. One further example might be, if we are already in debt, with a balance of −£10, and needed some more money, say another £10, our debt would rise to £20, i.e. (−10) + (−10) = −20. Most primary children do not go beyond this level when working with directed numbers.

Previous knowledge required

Read, write and count whole numbers. Is familiar with the use of a number line. Can record using graphs.

Resources needed for Lesson 1

Number lines (marked −50 to +50), cards numbered −50 to +50, dice marked 1–6.

Resources needed for Lesson 2

Copymaster 4, collection of weather reports from newspapers over a few weeks, large drawing of a thermometer.

Resources needed for Lesson 3

Number lines marked −50 to +50, cards marked −50 to +50, dice marked on each pair of opposite sides − and +, counters (different colours).

Teaching the lessons

Lesson 1 ①

Key questions

How many ways can you make a jump?
Which number did you start with and why?
Is more than one size jump possible?

Vocabulary

Negative number, count on, count back.

Introduction | 15min

Introduce the number line marked −50 to +50 to the children. Ask the children what the number line could represent. Talk about temperature, bank balances, and lifts. Ask the children to pick a number on the number line and then you choose a number. Ask the children if they can get to your number from their number in equal jumps of 2,3,5 ... Try a few more examples. Point out that the children can jump in either direction.

Activities | 30min

Make sure each pair of children has a number line, set of cards and a die. Explain to the children that they are going to carry out an activity similar to that in the introduction. Tell the children to shuffle the cards. Each child in the pair takes a card. One child in each pair throws the die. Both children show each other their numbers and mark them on the number line. They have to get from one number to the other number on the number line using the number of the set indicated by the number on the die. Remind the children they can move in either direction, the jumps must be of equal size and they must use whole numbers. Is it always possible? Discuss possible strategies. Tell the children to record their results and have as many turns as they can.

Closing the lesson | 15min

Ask the children which pairs were able to get from one number to the other. Ask for their results. Ask the children to explain their work. Finish the session as you started, asking individuals for their ideas.

Assessment

Child's performance	Teacher action
Can make jumps of one or two using positive numbers	More activities using the number line
Can make a variety of jumps	Give other examples of negative numbers
Uses both positive and negative numbers, and calculates jumps mentally	Move on to next lesson

Lesson 2 ②

Key questions

How far below/above zero?
What is the difference in temperature?
What do you have to add to −6 to get to −12?

Vocabulary

Zero, above zero, below zero, compare, positive number, negative number.

Introduction   10 min

Show the children the latest weather report. Read out the temperature. Ask a child to mark the temperature on the large thermometer drawing. Ask what would be the new temperature if it changed by 10, 15, −6, −12 ... degrees. Each time mark the new temperature.

Activities   35 min

Give the children **Copymaster 4** and copies of the weather reports over the last week or so. Tell the children to record the temperatures on the copymaster. Ask the children to work out how much the temperature has changed from one day to the next, and record their results. Make a graph of the results.

Closing the lesson   10 min

Ask the children to share their results. Finish the session by asking questions such as: if it is −6 degrees now and tonight it is 7 degrees colder, what will be the new temperature?

Assessment

Child's performance	Teacher action
Cannot use negative numbers	Give more experiences with the extended number line
Can locate new temperature but cannot calculate difference	Consider repeating Lesson 1
Locates and calculates difference	Move on to next lesson

Lesson 3 ③

Key questions

How many moves did you make?
How close to zero have you been?
What card do you need to get to zero?

Vocabulary

Zero, negative number, positive number.

Introduction   15 min

Review the last lesson. Talk about the change in temperature, the number line, moving up and down, zero, positive and negative numbers.

Activities   30 min

Make sure each child has a complete set of equipment. Tell the children they are going to play a game in pairs but they must each record their results and make a record of the moves they made. Tell the children to shuffle the cards. Children put their counters at zero. They take it in turn to pick a card and throw the dice. They move their counter the number of places shown on the card in the direction indicated by the dice, − left + right. Every time they land on zero they get 10 points. The winner in the pair is the one with the most points.

Closing the lesson   15 min

Collect the records of the children's games. Use the results of the game to pose some questions using directed numbers that the children have to calculate mentally.

Assessment

Child's performance	Teacher action
Unsure of where to place counter	Use the ideas in Lesson 1 again
Unsure of how to record game	Use the ideas in Lesson 2 again
Plays game successfully	The learning targets for this theme have been met

Homework

Children to make a list of all the machines in their houses which use temperature gauges. Keep a daily weather chart for a week. How does the temperature change? Record their pocket money expenditure and make comparisons with bank balances. (This last task to be used at your discretion.)

Number systems, symbols and words

Learning targets

On completion of this theme the children should be able to:

1 ➡➡ appreciate some key elements in the development of our number system

2 ➡➡ discuss the importance of place value in our system

3 ➡➡ explore language associated with number.

Before you start

Subject knowledge

Numbers and number systems have a history and that history is on a world-wide basis. Many societies in different places and at different times have taken leadership in mathematical thinking. The Babylonians, Ancient Greeks, Egyptians, Chinese, Hindus and those who developed the Arabic numbers we use have all had a formative influence on mathematics, and this influence can be seen to this day. For example, the Babylonians based their number work around 60 and we still use 60 seconds in a minute, 60 minutes in an hour, and 6×60 degrees in one complete rotation.

The key historical ideas for the children are the development and importance of place value, and the role of language in relation to mathematical signs and symbols. The Arabic numerals we use in Britain evolved into the form we know today. An important figure in their development was the Arabic mathematician al-Khwarizmi who promoted the system in a booklet published in AD 825. We still use his name today in the corruption 'algorithm'.

Previous knowledge required

Confident in all aspects of the four operations. Research skills including history and mathematics required.

Resources needed for Lesson 1

Copymaster 5, Copymaster 6, atlases, appropriate reference books, encyclopaedias, CD-ROM versions of such reference works if possible.

Resources needed for Lesson 2

Copymaster 7, atlases, appropriate reference books, encyclopaedias, CD-ROM versions of such reference works if possible, a clock face with Roman numerals might be useful.

Resources needed for Lesson 3

Copymaster 8, dictionaries, appropriate reference books, encyclopaedias, CD-ROM versions of such reference works if possible.

Teaching the lessons

Lesson 1 ①

Key questions

What base for counting is being used here?
Where did our numbers come from?
Why do you think number systems have evolved over the years?

Vocabulary

Base, counts, counting, number words.

Introduction [15 min]

▦ Tell the children that the symbols we use, 0 to 9, are an invention and they were developed over many centuries in other parts of the world. Point out that we call our numbers 'Arabic' because of their origins in the Middle East although Arabic mathematicians built on earlier work in the evolution of their number system. Tell the children that in this lesson they are going to look at some different number systems that were in use, and some that are still in use in some parts of the world. To help orientate them to the tasks that follow put the counting numbers from 0–20 on the board and discuss what, e.g. 'teen' means. Now do the same with numbers in French – how does this language work with numbers?

Activities [25 min]

⚙ Give out **Copymaster 5** on which there are examples of words for numbers from different parts of the world. The children have to try and work out how the systems work from these words. They also have to try and associate words that describe the same numbers.

As appropriate, give out **Copymaster 6** to small groups or pairs. Here the work is extended to finger multiplication – a device with a long history.

Closing the lesson [10 min]

With the whole class, concentrate a discussion on Copymaster 5. Explore what the children have discovered. In the first example, the Bushmen have a system based on 2s. The Aztecs had a system based on 5s. The table below sets out the words in the final section of Copymaster 5. If there is time then get someone to demonstrate some finger multiplication.

English	French	German	Latin
one	*un*	*ein*	*unus*
two	*deux*	*zwei*	*duo*
three	*trois*	*drei*	*tres*
four	*quatre*	*vier*	*quattuor*
five	*cinq*	*fünf*	*quinque*
ten	*dix*	*zehn*	*decem*

Assessment

Child's performance	Teacher action
Finds the task very challenging	Check on understanding of place value and the importance of 10 in our system
Completes most of Copymaster 5	Give more time whilst checking the clarity of the thinking about bases
Completes Copymaster 5 and makes a start on finger multiplication using Copymaster 6	Allow child to get further into finger multiplication and especially how it works

Lesson 2 ②

Key questions

What is the meaning of 'zero'?
Why is a place value system so important?

Vocabulary

Place value, zero and other number names.

Introduction [10 min]

Remind the children of what they know about the origins and characteristics of the Arabic number system, stressing its place value and the meaning of zero. Tell them that they are going to be finding out whether two other ancient systems are easier or harder to handle than the Arabic system.

Activities [25 min]

Give out **Copymaster 7**. If necessary, remind the children of some of the conventions in writing Roman numbers. A clock face, if available, might help.

Closing the lesson [15 min]

Using individuals, sample from the responses to Copymaster 7. Work through any misunderstandings on the board. Finish by pointing out that without a place value system you have to have more and more symbols – use the Romans as an example.

Assessment

Child's performance	Teacher action
Has difficulty in applying knowledge to the given systems	Place value is so central that it is worth going over this again
Makes steady progress with the odd false start	Give more time then check understanding
Copes well with all the key ideas	Involve child in developing a resource bank or display

Lesson 3 ③

Key questions

What does this mean?
How many in ...?

Vocabulary

'Prefix' may need some explanation in your introduction. The lesson is centred on vocabulary.

Introduction [15 min]

Ask the children for words they know which depict a certain quantity – if need be put up 'dozen' to start them off. Here are a range of possibilities.

Pair	2	Half a dozen	6
Couple	2	Dozen	12
Brace	2	Baker's dozen	13
Twins	2	Score	20
Triplets	3	Half century	50
Quadruplets	4	Century	100
Quintuplets	5	Gross	144
Sextuplets	6		

Activities [20 min]

Give out **Copymaster 8** and ask the children to use dictionaries and other reference works as necessary.

Closing the lesson [15 min]

Go through the first part of the copymaster briskly as these words should be familiar. Spend more time on the B examples the children come up with.

Assessment

Child's performance	Teacher action
Is not familiar with a number of the words	Spend time on developing the vocabulary: this will need to be over weeks rather than days
Has a reasonable knowledge of the vocabulary	Encourage further extension by using some of the less familiar words in subsequent mathematics sessions
Knows most of the words and contributes others	The learning targets for this theme have been met

Homework

The activities in this theme all lend themselves to individual extension. This could either be through additional examples or personal research by the children.

Investigations

- What is the biggest number you can think of? How do you know it's the biggest?
- How many ways can you make −10? Use a number line to help.
- What is the biggest ancient number symbol you can find? Can you write a bigger one?
- How fast can you count on in 2s to 101 starting with 1?
- How long does it take your heart to beat 10 times? What about 100, 1000, 1 million times?
- How high are ten 50p pieces? What about 100, 1000 and so on? Taller than the tallest building?
- Use a blank number line to produce a time-line in decades. Put in the year you were born then find three facts for each decade on your line.
- Investigate consecutive numbers: what sort of product do they give? What about adding consecutive odd numbers: what sort of numbers are produced?

- TV and films use Roman numerals for the year so, e.g. the year 2000 will be written MM. Explore other years, birthdays, anniversaries, ages and so on. Make a collection of favourite films/programmes and when they were made.
- Make up some problems using Roman or Egyptian symbols. Try all four operations.
- A group uses a base 5 system in which 5 units = 1 finger, 5 fingers = 1 hand, and 5 hands = 1 arm. How would you write 10, 100, 1000 in such a system? Can you add and subtract?
- How many years in a decade, century, millennium? What is the difference between each? How many decades in a century and a millennium? How many centuries in a millennium?
- Try making a number square (like a 100 square) of the Egyptian, Roman, Babylonian or other ancient number system. Can you make some simple addition or multiplication squares using one or more of these systems?

Assessment

- Using place mats and sets of cards 0–9 make the biggest/smallest numbers you can. Take turns with a partner. Strategy will come into this. Each time you have the bigger number you get 10 points. First to 100 points wins.
- Give the children a number line for 0–1000 and ask them to choose a number between 100 and 200. Then they have to ring those numbers on the number line with the same number of tens and units, but different hundreds, as the number they have chosen.

- Use a 0–100 number line. Such a line can be made up from Copymaster 63. Ask the children to choose a number and then see how many jumps of 10 before they reach 100. Do the same with 0–1000 but jumping in 100s.
- Using a −10 to +20 number line how many ways can they make 10?
- 399 + 1: what is the answer? What about 3999 + 1? Try to invent others like this.

ADDITION AND SUBTRACTION

There are a number of key characteristics that children who can use addition and subtraction exhibit. The first is a good understanding of place value.

Place value underlines all aspects of addition and subtraction, whether it is with whole numbers or decimal numbers. The most common mistakes that children make reflect their lack of understanding of place value. This is especially true when using formal algorithms which involve carrying a number to the next column to the left, when children are frequently unsure of which number to carry. Other common errors are: putting values in the wrong column, i.e. placing a 10 in the units column; and being unsure of which digits are added together, resorting to the strategy of adding all the digits regardless of value.

When children move beyond whole numbers to using and calculating with decimal numbers the problems are frequently compounded. This can clearly be seen in adults' failure to understand how to add and subtract (and also multiply and divide) decimal numbers. When working with decimal numbers children who do not understand place value resort to ignoring the decimal point and treat the decimal number as a whole number. Evidence from research indicates that it is not until the age of 15 that the majority of children overcome their misunderstanding of decimal numbers linked to place value.

Errors are not remedied by giving the child more examples to work through but by supporting and developing the child's knowledge and understanding of place value.

We also need to be careful with the examples and language we use when working with children. Often it is the 'rule of thumb' that we provide that later leads to problems when the child inappropriately applies the rule.

The providing of 'rules' for children is frequently part and parcel of an approach that starts the teaching of an algorithm with a simple example and then asks the children to work with more complex examples. The problem with this approach is that the children can frequently solve the simple examples intuitively, without understanding the underlying process. As teachers, we must make sure that children understand the concepts and principles when first teaching a topic such as addition and subtraction.

Children should also be encouraged to use their own informal methods of calculation. There is a great deal of evidence to show that children who develop a confidence with number have developed their own mental methods as well as knowing number facts 'by heart'. The knowing 'by heart' and informal mental methods actually support each other in the child's growing awareness of number.

Addition and subtraction to 10

Learning targets

On completion of this theme the children should be able to:

1 ➼ add and subtract to 10
2 ➼ add and subtract to 10 using a number line
3 ➼ relate addition and subtraction processes using numbers to 10.

Before you start

Subject knowledge

Addition and subtraction at this level is primarily about learning the number bonds to 10, as well as understanding order, i.e. 3 + 4 = 7, 4 + 3 = 7 does not matter for addition but does for subtraction. Children should encounter addition as the combining of two sets on the number line. Subtraction is more than just 'take away': the difference or comparison of two sets is equally important, as is the understanding that subtraction is the inverse of addition.

Previous knowledge required

A knowledge of all numbers to 10. The ability to count to 10. An understanding of some of the language associated with early addition and subtraction, e.g. more, fewer, add, take away, etc. The skill of writing numbers and the ability to record.

Resources needed for Lesson 1

Copymaster 9, spinners made up from Copymaster 61, paper and pencil or books for recording, counters (two different colours).

Resources needed for Lesson 2

Copymaster 62, blank dice numbered 0–5 (and other blank dice available for extension opportunities).

Resources needed for Lesson 3

Copymaster 10.

Teaching the lessons

Lesson 1 ❶

Key questions

Which numbers do you need to cover to win the game? Which numbers do you need your spinners to show to make 9, 7 …?
What is the difference between the numbers on the spinner?

Vocabulary

Add, altogether, more, how many more, take away, difference between, less.

Introduction 20 min

▦ Ask the children if they can suggest different ways of making 5. Ask the children to do the same for 10, suggesting they can use more than 2 numbers if they wish. Record their solutions on a chart. Ask the children if they can see any patterns.

Activities 25 min

▦▦ Explain to the children that they are going to work in pairs playing a game. Show the class **Copymaster 9** and the spinners. Explain the following rules.

1 Each pair chooses two different colours for their counters.
2 The object of the game is to cover four numbers on the copymaster in a row.
3 Each child takes it in turn to spin both spinners.
4 The number for each spinner may be added together or children can choose to use them separately. The more able child should also use the difference between the two numbers to cover a square.

The first person to cover four squares in a row, that is vertically, horizontally or diagonally, is the winner. The children should record what operations they used, if any, to cover the squares.

Closing the lesson 15 min

▦ Bring the group together. Ask the children to explain how they made the numbers they covered on the copymaster. If a child offers one solution, ask if there were any other ways of making the same number. Ask the children if anyone had used the difference between two numbers. Record the children's results for everyone to see and ask if there are any patterns. Review the work using the key questions.

Assessment

Child's performance	Teacher action
Only uses numbers shown on the spinners	Give more experiences of counting and adding with apparatus
Can use addition	Move on to next lesson
Subtraction, the difference between two numbers used	Move on to next lesson

Lesson 2 ②

Key questions

How many ways to 10?
How many ways to 0?
Can you use more than 2 numbers?

Vocabulary

Pattern, difference between.

Introduction | 20 min |

 Show the children the number line marked 0–10. Throw a die and ask the children to explain how many different ways they can get from the number on the die to 10. Mark the different ways on the number line. Ask the children how many ways they can get from the number on the die to 0. Mark them on the number line.

Activities | 25 min |

Explain to the children that they are going to use the dice and number line to explore the different ways of making and 'unmaking' numbers. Give the children a number line (**Copymaster 62**) and two dice each marked 0–5. Ask the children to investigate the different ways of making numbers up to 10 and 'unmaking' them to 0. Tell the children to throw the 2 dice and add the numbers. Mark the result on the number line. How many ways can they find to get to 10? How many ways to 0? Support the children by asking them to explain their results. The more confident child might use 3 dice (1 of which is marked 6–11) and a number line marked 0–20.

Closing the lesson | 15 min |

Bring all the children together. Ask some of them to show their results and explain any patterns they found. Finish the lesson by having a quiz. You call out a number between 0 and 10 and each child must call out a way of making the number, using 2, 3 or more numbers.

Assessment

Child's performance	Teacher action
Only adds the thrown numbers	More experience of adding pairs or trios to make 10
Can only make numbers up to 10	More experience of using subtraction to undo addition
Can complete task	Move on to Lesson 3

Lesson 3 ③

Key questions

What number do you think is missing?
Is there any pattern?

Vocabulary

How many more to make …? Total, add, sum, difference between, equals, leaves, subtracts.

Introduction | 20 min |

 Ask the children how many ways they can make 5, 6, 9 … Record their ideas so everyone can see the results. Ask the children how many different ways they can find to make 9 using addition and subtraction. Discuss with the children the different results; look for patterns in their ideas.

Activities | 25 min |

Show the children **Copymaster 10**. Explain that someone in the class has spilt paint over the resource sheet, and they are going to have to try and discover the missing numbers. Move round the groups giving support and asking the children to explain how they arrived at their solutions.

Closing the lesson | 15 min |

Bring the groups together and ask for a volunteer from each one to explain how they solved the problems.

Assessment

Child's performance	Teacher action
Can only solve some of the addition problems	Needs further experience of addition and subtraction
Can solve most of the addition and subtraction problems	The learning targets for this theme have been met
Can solve all the problems	The learning targets for this theme have been met

Homework

Suggest to the children that they draw a piece of wooden furniture they have at home. Ask the children to count the number of straight pieces, number of curved pieces, and how many pieces altogether. Ask the children to count the number of letters in their name. Ask the children to count the number of letters in the names of family members. Ask them to find out who has the most, and least, and what is the difference?

Addition and subtraction to 20

Learning targets

On completion of this theme the children should be able to:

1 ➡➡ recall and use number bonds to 20
2 ➡➡ recognise addition and subtraction facts to 20
3 ➡➡ add and subtract to 20.

Before you start

Subject knowledge

The use of number bonds and strategies for calculation such as 'near doubles' should be stressed at this level. Addition is both the adding of 2 sets and 'counting on' using the number line. Representing subtraction on the number line by 'counting back' makes a link with addition and provides a different model from 'take away'. The numbers 10 to 20 are important because they provide the pattern for operations using higher numbers that children meet later, e.g. 0–100, etc.

Previous knowledge required

The ability to read, write and count numbers to 20. Can recall the number bonds to 10 and has experience of adding and subtracting to 10.

Resources needed for Lesson 1

Copymaster 62, dice (variety of different types, e.g. with 6, 8, 12 faces; faces numbered 11 and 12 masked), felt-tipped pens.

Resources needed for Lesson 2

Copymaster 11.

Resources needed for Lesson 3

Calculators.

Teaching the lessons

Lesson 1 ①

Key questions

What is the double of …?
What is 1 more than the double of 6, 5, 7 …?

Vocabulary

Double, more, add, total, odd, even.

Introduction [20min]

▨ Ask the children if they know what is meant by double; talk about doubles, giving some examples. Ask the children what the double of 6 is and 7, 8, 4 …? Write down the different responses from the children, encouraging them to look for a pattern. Ask the children what they think a near double is. Talk about near doubles being one more or one less than a double. Ask the children to add two numbers that are next to each other (e.g. 8 + 9), tell the children that knowing the double of 8 or 9 would help with the calculation, and ask them to explain how.

Activities [25min]

▨ Explain to the children they are going to look for doubles using the different dice and a number line

▣▣ (**Copymaster 62**). Ask the children to throw a die and find the double on the number line. Tell the children to find as many doubles as they can, using the dice and number line. After 15 minutes or so, ask the children to find and record all the near doubles for the doubles they have already found. Ask for an explanation of how they worked out the doubles. Some children might talk about the 2× table; encourage them to remember the number facts.

Closing the lesson [15min]

▨ Ask the children specific questions: what is the double of 6, 4, 9 …? Ask the children to calculate addition sums using the near double strategy.

Assessment

Child's performance	Teacher action
Cannot find many doubles	More practice and experience of doubles needed
Can find some doubles but unsure of near doubles	Provide activities using odd and even numbers
Can find doubles and uses the near double strategy with confidence	Move on to next lesson

14

Lesson 2 ②

Key questions

How many possibilities did you find?
Did you find any patterns in your results?

Vocabulary

More, less, difference, add, subtract, pattern, minus.

Introduction 15min

Prepare the session by drawing a large, but simple, dartboard (see **Copymaster 11**). Tell the children they are going to play a special game of darts where the scores for the darts are summed. Explain that they each have three darts, and for any given score they have to work out where the darts might have landed. Ask how they might score, e.g. 17, 12, 9 ....

Activities 35 min

Give out Copymaster 11. Ask the children to choose a score on the dartboard and try to find all the possibilities of making that score with three darts. Get the children to suggest ways of recording their ideas. After 5 or 10 minutes ask the class as a whole for some of their ideas/results. Pose the question: if two or more darts landed on the same number, how would that affect the possibilities that you have found? Let the children investigate this problem for a while then ask if they can find any pattern to their results. The children can then try using their dart scores as subtractions. Thus they can begin with 20 and see which three scores could bring them to 0.

Closing the lesson 10min

Bring all the children together. Review the lesson by asking some of the key questions. Ask the children to show and explain their results. Ask the children specific questions: If I scored 13, 16… where could my darts have landed? If I scored 18 and one dart landed on 7 which numbers could my darts have landed on? If I scored 20 which numbers could my 3 darts have landed on? If I start with 20 how can I reach 0 by subtracting my 3 scores?

Assessment

Child's performance	Teacher action
Has difficulty in finding possible results	Provide more experiences on doubles, revisit Lesson 1
Chooses to work on low numbers, has difficulty working with numbers beyond 10	Give experiences that build on number facts to 20
Has no difficulty with the task	Move on to next lesson

Lesson 3 ③

Key questions

What numbers can you make?
Can you explain how you made your number?
Can you find another way of making the number?

Vocabulary

Equals, minus, add, plus, subtract, difference, how many?, less, sign, leaves, total.

Introduction 15min

Introduce the children to the calculator. Tell the children they are going to pretend that the calculator is broken and they can only use certain keys, 1, 2, 3, 4, +, −, =. Make sure the children understand this before going any further. Tell the children if they wanted to write 15 using the broken calculator they could use $4 + 4 + 4 + 3 = 15$ or $12 + 4 − 1 = 15$. Ask the children for their ideas and record them so everyone can see the results. Choose another number; ask the children to suggest possibilities.

Activities 35 min

Tell the children they are going to try and write all the numbers to 20 using the broken calculator. Ask the children to record their results, looking for patterns and as many results for each number as possible. Support the children by asking them to explain their results. Suggest ideas such as repeated subtraction as well as repeated addition.

Closing the lesson 10min

Review the lesson by asking the children to explain their results. Give the class a challenge of suggesting possible solutions as quickly as they can for a given number.

Assessment

Child's performance	Teacher action
Has difficulty with the task, only uses addition	More experiences of number facts; try Lessons 1 and 2 again after a suitable period
Can find at least one solution for most numbers	More activities that look at subtraction facts to 20
Finds more that one solution to all the numbers and uses both subtraction and addition	The learning targets for this theme have been met

Homework

Write down the ages of all the people aged under 18 in your family (include brothers, sisters, cousins, etc.). 'Play around' with these numbers making as many totals under 20 as you can, using addition and subtraction.

Addition of numbers to 100

Learning targets

On completion of this theme the children should be able to:

1 ➡➡ model the addition of two-digit numbers on the number line

2 ➡➡ model the addition of two-digit numbers using base 10 materials

3 ➡➡ solve two-digit addition problems.

Before you start

Subject knowledge

In developing an understanding of addition beyond the recall of simple number bonds there are 2 key ideas. Firstly, place value – that is the value of a digit is determined by its place. Secondly, exchanging – the 10 in one column can be 'exchanged' for 1 in the next column to the left. While it is important that children learn and understand the formal algorithm for addition, it is equally important that they are encouraged to use their own informal methods of calculation. The use of informal methods gives recognition to the fact that the way we calculate mentally differs from the paper and pencil algorithm, and that the informal supports the formal.

Previous knowledge required

Can read, write, count and order numbers to 100 and has the ability to count in 10s. Can recall and use number facts to at least 20. Can add and subtract to 20.

Resources needed for Lesson 1

Copymaster 63, a large number line marked 0–100, felt-tipped pens, sets of cards made up from Copymaster 64.

Resources needed for Lesson 2

Copymaster 63, base 10 materials (or Multi-link®), Copymaster 64.

Resources needed for Lesson 3

Base 10 materials (or Multi-link®), Copymaster 63, Copymaster 12.

Teaching the lessons

Lesson 1 ①

Key questions

How many altogether?
How many jumps of 10, of 1 …?

Vocabulary

Count on, digit, two-digit number, tens, units or ones, altogether, jumps.

Introduction [15 min]

▦ Introduce the children to the large number line. Ask the children how they would show 27 on the number line, counting in 10s and units. Draw the 2 ten jumps and 7 single jumps on the number line. Ask the children how many 10s are in 27? How many units? Try a variety of different numbers making sure the children understand that two-digit numbers are made up of 10s and units. Tell the class that 2 children are saving to buy a toy car: one child has 23 pence the other has 34 pence. Ask the children if they can point out jumps on the number line that show how much they have altogether. Try other similar examples.

Activities [35 min]

👥 Show the children the cards numbered 0–50 (from **Copymaster 64**) and the number lines (from **Copymaster 63**). Shuffle the cards and ask a child to pick 2 cards, then another child to pick 2 cards. Ask the children each to add their 2 numbers using the number line, and ask who gets further from 0. Ask the children to pick another 2 cards each and ask who they think will get further from 0. The rest of the children work on this activity with their own sets of cards and number lines, trying as many examples as possible. Ask the children to record their work. Can they see any patterns?

Closing the lesson [10 min]

▦ Ask the children to explain some of their results. Did they find any patterns? What numbers made 70, 34, etc.?

Assessment

Child's performance	Teacher action
Has difficulty in showing numbers on number line	Give experiences that involve breaking two-digit numbers into 10s and units
Can show numbers on number line but has some difficulty with addition	More practice using the number line
Completes the task	Move on to next lesson

Lesson 2 ②

Key questions

Can you make 10?
Can you exchange your units for a 10?
How many 10s?
How many units?

Vocabulary

Tens, units, carry, exchange, digit, two-digit number, add, plus, sum.

Introduction   15min

Introduce the children to the base 10 materials. Explain the singles represent units or 1s and the longs, 10. Ask a child to suggest a two-digit number, then ask another child to make the number up using the base 10 materials. Ask for a few more suggestions and repeat the process. Now ask the children to suggest 2 two-digit numbers. Ask how they can use the materials to add the 2 numbers together. Carry out the addition, talking about making up 10 and exchanging 10 units for one long. Try this a few more times.

Activities   35 min

Make sure each pair of children has one set of cards (Copymaster 64), a number line (Copymaster 63) and base 10 materials. Remind the class of the Lesson 1 activity, where they each took a pair of cards and tried to find who would get further from 0. Tell the class they are going to carry out a similar activity. Explain they will be using the same two numbers, but only one child will be using the number line, and the other the base 10 materials to find the total. Ask the children to compare results. Are they the same? If not, what have they done? Ask the children to record their work.

Closing the lesson   10min

Talk about carrying and exchanging when using base 10 materials, and setting out numbers in columns. Give examples. To finish ask the children if they can recall the total of 34 + 53, 25 + 21…

Assessment

Child's performance	Teacher action
Can use the number line but not the base 10 materials	Provide more experiences using base 10 materials, grouping/counting in 10s
Can use both, but makes errors when exchanging	Give more activities involving carrying/exchanging
Is successful at both aspects of the activity	Move on to next lesson

Lesson 3 ③

Key questions

What do you have to add to make …?
How many units do you need? Do you think they have made a 10?
How many 10s do you need?

Vocabulary

Tens, units, columns, place value, group, count on, make.

Introduction   15min

Remind the children of the previous lesson, using base 10 materials to show some addition problems, e.g. if David has 26 marbles and Frances has 34 marbles, how many have they got altogether? Carry out similar 'real life' problems using a number line.

Activities   35 min

 Show the children **Copymaster 12** and explain that someone has spilt paint over some of the work. They have to work out the missing numbers. They can use base 10 materials or the number line (Copymaster 63) if they wish. Ask the children once they have been working for a while, if it is possible to find more than one solution for some of the problems. Remind the children about carrying and exchanging. Encourage the children to physically carry out the addition using apparatus if they need support.

Closing the lesson   10min

Ask the children to explain some of their solutions. Ask if there is more than one solution. Is there any pattern to the results and can they explain it? Encourage the children to share their ideas.

Assessment

Child's performance	Teacher action
Cannot solve many of the problems	More experiences of using base 10 materials and grouping, showing numbers on the number line in jumps of 10s and units
Can find solutions with the help of apparatus	Confidence building experiences using base 10 materials on similar problems
Confident, worked with the aid of apparatus	Progress on to the use of three-digit numbers

Homework

Use the digits in your telephone number to make pairs of two-digit numbers and find all the possible totals.

Subtraction of numbers to 100

Learning targets

On completion of this theme the children should be able to:

1 ➤➤ recognise that subtraction is the inverse of addition
2 ➤➤ use mental strategies to subtract two-digit numbers
3 ➤➤ subtract two-digit numbers.

Before you start

Subject knowledge

Subtracting using numbers greater than 20 demands that the children have an understanding of place value and decomposition. Children should have experiences of using base 10 apparatus, exchanging/decomposing numbers into 10s and units. Zero often poses problems in calculations – children need to be provided with plenty of practice where 0 is a digit in the first and second number as well as the answer.

Previous knowledge required

Can read, write, count and order numbers to 100.

Able to use and recall subtraction facts to 20. Has used a number line to count back.

Resources needed for Lesson 1

Copymaster 13, cards marked 1–100 (these can be made up using Copymasters 64 and 65).

Resources needed for Lesson 2

Sets of cards marked 1–100 (as for Lesson 1), number lines (there is a number line to 100 on Copymaster 63), 20 cards marked with subtraction problems using two-digit numbers.

Resources needed for Lesson 3

Base 10 materials, place mats, sets of problem cards with subtractions using two-digit numbers (10 of the problems should require decomposition).

Teaching the lessons

Lesson 1 ①

Key questions

How you get back to the first number?
How many ways can you turn the numbers round?

Vocabulary

Difference, inverse, subtraction.

Introduction ⌗ 15 min

Introduce the idea of the inverse by asking the children questions similar to: 24 + 7 = 31, how can I get back to 24 from 31? Try a few more problems, using increasingly higher numbers. Extend the idea of inverse to 'I am thinking of a number problem …'. Tell the children you are thinking of a number, e.g. you have added 4 then subtracted 6 and the answer is 8. So what was the original number? 8 + 6 = 14, 14 − 4 = 10 (10 was the original number). Try a few more examples. Make sure the children understand the idea of inverse.

Activities ⌗ 30 min

Give out sets of cards and **Copymaster 13**. Write on the board a calculation similar to 15 + 8 = 23, and tell the children that it is possible to use these same numbers to make another calculation, 23 − 15 = 8. Ask if they can suggest another way (23 − 8 = 15).

Give the children another example, 45 + 23 = 68, asking how they would re-write it. Try as many as you feel necessary. Tell the children to look at Copymaster 13; explain they are to shuffle the cards and then take 2 cards. They must try and make 3 different calculations then re-write them as they have just done.

Closing the lesson ⌗ 15 min

Talk about turning the addition around, telling the children that subtraction is the inverse of addition. Explain that subtraction 'undoes' addition. Finish the lesson by giving the children an addition problem and ask them to undo it, calculating mentally.

Assessment

Child's performance	Teacher action
Can only work with numbers less than 20	Number bonds, patterns to 30, 40 …
Can work with numbers greater than 20 but does not find all the combinations.	Work on similar tasks looking for patterns, number bond work
Finds all possible combinations even when using numbers greater than 50	Move on to next lesson

Lesson 2 ②

Key questions

How many 10s do you have to add?
How many units do you have to add?

Vocabulary

Difference, counting on, tens, units.

Introduction ⏱ 15min

🁢 Remind the class of Lesson 1, where they worked on the idea that subtraction is the inverse of addition. Tell the children that in this lesson they are going to learn how to subtract by adding. Tell the children it is sometimes called the grocers' method (its formal name is *complementary addition*). Give the children an example of a subtraction problem, $42 - 28$. Ask the children 28 plus what makes 42? Ask how many more 10s? Write on the board $28 + 10 = 38$, now ask how many more units? $38 + 4 = 42$, so the answer to $42 - 28 = 14$, because $10 + 4 = 14$. You have subtracted by adding, finding the difference between the two numbers. Try a few more examples.

Activities ⏱ 30min

👥 Give out the cards and number lines. Tell the children they are going to use the method you have just explained. They must shuffle the cards, take 2, and subtract the smaller number from the bigger by adding on. Remind the children first to think of the number of 10s they need, then units. Tell the children if possible to carry out the calculation mentally, then check their result by using the number line to find the difference between the 2 numbers by counting on. Try as many examples as possible. Does it always work?

Closing the lesson ⏱ 15min

🁢 Ask the children to give some examples of their work and explain what they found out. Finish the session by using the cards with the subtraction problems. Show the children a card for a few moments. They must try and use the method they have just worked on to give solutions.

Assessment

Child's performance	Teacher action
Gets confused with the 10s and units	Number line activities counting on and back in 10s and units
Cannot calculate mentally	Work on number bonds, quick fire number games
Calculates mentally	Move on to next lesson

Lesson 3 ③

Key questions

How many units/10s?
What could we use here?

Vocabulary

Difference, counting on, tens, units, subtract, subtraction.

Introduction ⏱ 25min

🁢 Remind the children of the previous lesson with some questions written on cards. Show the cards only for a few moments; then ask for the answers. Set out a subtraction problem formally in columns using the base 10 materials and the place mat as shown here.

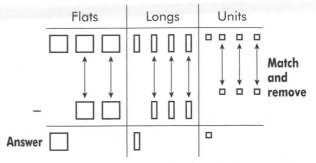

Use numbers that do not require decomposition. Physically carry out the subtraction by first matching each unit piece from the lower number to a piece in the upper number and remove. Any excess unit pieces remain. Carry out the same procedure with the 10s column. The number of 10s and units remaining is the answer. Work out a few more examples. Introduce decomposition by making the lower number in the units column bigger than the upper number. Carry out the subtraction by first matching the units. When you 'discover' you do not have enough units in the lower number to match with those in the upper number ask the children what should be done. Suggest exchanging one of the 10s from the 10s column. Physically exchange the long for 10 units, then carry on matching and removing the units. Carry out the matching and removing in the 10s column. Try a few more examples. Write out an example on the board and formally work through the process, reminding the children of what they have just seen when using the base 10 materials.

Activities ⏱ 25min

👥 Give out the base 10 materials, the cards and the number lines. Tell the children to work through the problem cards, writing out their work using the base 10 materials. They may check their answers using the number line by counting on.

Closing the lesson ⏱ 5min

🁢 Finish the lesson by going over some of the subtraction problems formally on the board.

Assessment

Child's performance	Teacher action
Cannot subtract two-digit numbers	Practical activities using objects and number lines
Can subtract two-digit numbers but cannot do decomposition	Practical activities using base 10 materials
Can use decomposition	Move on to three-digit and greater numbers

Homework

Find out how old your grandparents are. Subtract your age from theirs, to find the difference. How many ways can you reach 50 from 100 using only subtraction?

Addition of numbers greater than 100

Learning targets

On completion of this theme the children should be able to:

1 ➥ estimate answers to the addition of three-digit numbers
2 ➥ add three-digit numbers on the number line
3 ➥ recognise addition patterns to 1000.

Before you start

Subject knowledge

The addition of numbers over 100 builds on the experiences and knowledge that children have already acquired working on two-digit numbers. The key idea of place value is still important and children need to be reminded of its importance. The idea of exchanging and moving to the left every time that a group of 10 is made should be reinforced at this stage and later. Children should be taught how to estimate the answers to problems by rounding up and down, again linked to place value, rounding to the nearest 10, 100 ...

Previous knowledge required

Ability to use and recall number bonds to 100. The ability to add two-digit numbers. An understanding of place value.

Resources needed for Lesson 1

Copymaster 14, set of cards with three-digit numbers written on them.

Resources needed for Lesson 2

Number line marked 0–1000, base 10 materials, cards marked 0–9 (Copymaster 66).

Resources needed for Lesson 3

Calculators.

Teaching the lessons

Lesson 1   ①

Key questions

How close is the number to the nearest 10, 100 ...?
Do you round up or down?
What is your estimate?

Vocabulary

Round up, round down, tens, units, hundreds.

Introduction |15min|

Write some three-digit numbers on the board: 234, 567, 123, 678 ... Ask the children how they can make these numbers easier to add together. Tell the children you do not want the actual answer to the calculations; only an approximate answer. Talk about rounding up and rounding down numbers. Point out to the children that 234 is near to 230 and 567 is near to 570 and adding 230 and 570 is much easier. 230 + 570 = 800, so the answer to 234 + 567 must be around 800.

Talk about rounding numbers to the nearest 10: if the last number is below 5 round down; if 5 or above

round up. Write some numbers on the board and ask the children to mentally round up or down the numbers to the nearest 10. Ask the children how they would round the same numbers to the nearest 100: 234 = 200, 567 = 600, 123 = 100 and 678 = 700. Ask the children to explain what you have done. Try some more numbers, rounding to the nearest 100.

Activities |30min|

Give the children **Copymaster 14**. Explain to the children that they must round the numbers to 10 then to 100 and work out the answers before determining the actual answer. Allow the children to use any method they prefer when they carry out the actual addition.

Closing the lesson |15min|

Ask the children to share their estimates, and compare and discuss these with them. How accurate are the approximations? Which are the most useful when compared with the actual solutions? Hold up the cards with the three-digit numbers written on them, asking the children to round the numbers to the nearest 10, 100 as quickly as possible.

Assessment

Child's performance	Teacher action
Finds it difficult to round numbers	Give more place value activities
Can round to 10, has some problems with three-digit numbers	More experiences of rounding three-digit numbers
Rounds three-digit numbers to 10, 100.	Move on to next lesson

Lesson 2 ②

Key questions

How many jumps of 100s, 10s, 1s?
Which number is it better to start from?
How many more 10s, hundreds …?

Vocabulary

Hundreds, tens, units, count on, count back.

Introduction | 15min |

 Introduce the 0–1000 number line to the children. Ask the children to suggest a three-digit number; mark it on the number line. You choose a bigger three-digit number and mark it on the number line. Ask the children: how many jumps is the second number more than the first? Try a few more numbers. Ask the children questions similar to: what is 45 plus 467? Ask the children which way they would do this on the number line and request a volunteer to carry out the addition.

Activities | 30min |

 Give each pair of children a number line, some base 10 materials and a set of cards. Remind the children of how they made numbers in the last lesson. Ask them to shuffle the cards, take 2 and make as many numbers as possible. Both children add the same 2 numbers but one uses the base 10 materials and the other the number line. They should compare results. They then repeat the process, for all the number cards. Ask the children to record their work.

Closing the lesson | 15min |

 Allow the children to share their work, explaining their results and how they carried out the calculations. Finish the lesson by using the number line to carry out some more additions.

Assessment

Child's performance	Teacher action
Unsure of using the 0–1000 number line	More number line activities, maybe going back to the 0–100 line
Has some difficulty adding three-digit numbers	Activities using base 10 materials
Adds three-digit numbers	Move on to next lesson

Lesson 3

Key questions

How many equal jumps?
Can you find another way to make a jump?

Vocabulary

Jumps, tens, hundreds, thousand.

Introduction | 15min |

Start the lesson by using a number line to give the children some addition problems similar to: I am on 543, how many jumps to 843? I am on 300, how many jumps to 1000? I am on 345, how many jumps to 1000? Allow the children to decide how they 'organise' their jumps – in 10s, 100s, units.

Activities | 30min |

Give each child a number line or calculator. Tell the children they must choose a number. Then ask them to find as many ways as possible to get from their chosen number to 1000 in equal jumps. If they cannot find many tell the children to choose another number.

Closing the lesson | 15min |

Ask for an explanation of any patterns they may have found. Let the children explain their results. Ask the children how many ways they can think of to get from 250 to 1000 in equal jumps. Finish the lesson by asking a number of similar questions.

Assessment

Child's performance	Teacher action
Can get to 1000 but only when starting from 200, 400 …	Limit the number to reach, e.g. 200
Can get to 1000 starting from numbers similar to 250 …	Use calculators to explore ways of making 1000
Can get to 1000 using a range of different starting positions and in a variety of jumps	The learning targets for this theme have been met

Homework

Write down any 2 numbers with 3 digits each. Add them together. Then add the numbers in the answer together and continue adding the digits until only a single digit remains. Now alter the order of the digits in the starting numbers and try again. Do you always end up with the same final digit?

Subtraction of numbers greater than 100

Learning targets

On completion of this theme the children should be able to:

1 ➡➤ subtract three-digit numbers using base 10 materials
2 ➡➤ find number combinations and subtract three-digit numbers
3 ➡➤ investigate subtraction problems.

Before you start

Subject knowledge

The use of the formal algorithm for decomposition presents some children with problems when using three-digit numbers, especially if there are 0s in the larger number. Typically, when presented with a problem such as 300 − 236, they are unsure where to start the calculation. In part, this is because they are insecure in their knowledge of place value, but they also have not really understood decomposition.

Previous knowledge required

Knows number bonds to 100. Can subtract two-digit numbers. Has used base 10 materials before.

Resources needed for Lesson 1

Copymaster 15, base 10 materials.

Resources needed for Lesson 2

Packs of cards made up from Copymaster 66.

Resources needed for Lesson 3

Copymaster 16.

Teaching the lessons

Lesson 1 ①

Key questions

Which can you subtract – the units or the 10s?
Can you exchange a long for 10 units?
Can you exchange a flat for 10 longs?

Vocabulary

Exchange, column, units, longs, flats, tens hundreds.

Introduction 20 min

▦ Write a three-digit subtraction problem on the board, e.g. 263 − 185. Ask the children how they would attempt the problem. Discuss all their ideas. Tell the children you are going to show them a method using base 10 materials. Remind the children of work they have carried out using two-digit numbers. Start with the problem on the board, laying out 2 flats, 6 longs and 3 units, and 1 flat, 8 longs and 5 units. Tell the children to look at the units column, pointing out that you need to take a long from the 10s column

and exchange it for 10 units before you can carry on. Now you have 13 units subtract 5 which leaves 8. Tell the children you now have only 5 longs and you have to subtract 8. Tell them that you will have to exchange a flat for 10 longs. Now you have 15 longs, show the children, then take 8 away leaving 7. Show the children you now have only 1 flat remaining, from which you must subtract 1 flat. Ask the children what the answer to the problem is (263 − 185 = 78).

Try a few more examples working through the problem with the children.

Activities 25 min

👤 Give out **Copymaster 15**. Point out that the first calculation they tried with you is on the copymaster as an example. Tell the children how to carry out the tasks on the copymaster, using base 10 materials if they need to.

Closing the lesson 15 min

▦ End the lesson by carrying out decomposition (as you started the lesson), getting the children to help in explaining what you have to do.

Assessment

Child's performance	Teacher action
Is unsure where to begin the calculation	Go back to subtracting to 20
Can subtract but unsure about the 0	More practical activities using base 10 materials
Can subtract three-digit numbers	Move on to next lesson

Lesson 2 ②

Key questions

How many numbers can you make?
What is the value of …?
How did you find the difference?

Vocabulary

Difference, value, combination, tens, units, hundreds.

Introduction [15 min]

Remind the children of the last lesson. Work through a decomposition problem with the whole class as in the last lesson. Make sure the children are involved by asking questions all the time.

Activities [30 min]

Give out the packs of cards. Tell the children to shuffle the cards. Tell each child they must take 3 cards. You take 3 cards, show them to the children and then write the digits on the board. Explain you are going to find as many different three-digit numbers as possible with your three digits, e.g. 3 5 7 gives 357, 375, 573, 537, 753 and 735. Tell the children once they have found all the numbers they must find the difference between the greatest number and the smallest number. Once they have done that they take another 3 cards.

Closing the lesson [15 min]

Ask the children to share some of the number patterns they found. Ask the value of the different digits. Ask the children to explain how they calculated the difference. Finish the lesson writing 4 numbers on the board and challenging the class to find all the possible combinations.

Assessment

Child's performance	Teacher action
Cannot find all the possible numbers	Number pattern activities
Cannot subtract	Repeat activities similar to Lesson 1
Finds all possible numbers and subtracts	Move on to next lesson

---

Lesson 3 ③

Key questions

Where would be a good place to use the 0?
Can you find a number where you don't have to exchange?
Do you have to exchange in the units, 10s …?

Vocabulary

Difference, subtract, units, tens, hundreds, thousands, exchange.

Introduction [10 min]

Review the last lesson, reminding the children of how to subtract three-digit numbers. Ask the children to work out a few examples.

Activities [35 min]

Give out **Copymaster 16** to each child. Explain that they use all the digits 0–9, once and once only, to fill in the missing numbers. Here are some examples.

Examples of subtraction problems using digits 0–9 only once each

$$
\begin{array}{r} 1035 \\ -\ 789 \\ \hline 246 \end{array}
\qquad
\begin{array}{r} 1026 \\ -\ 539 \\ \hline 487 \end{array}
\qquad
\begin{array}{r} 1503 \\ -\ 829 \\ \hline 674 \end{array}
$$

$$
\begin{array}{r} 1035 \\ -\ 746 \\ \hline 289 \end{array}
\qquad
\begin{array}{r} 1062 \\ -\ 483 \\ \hline 579 \end{array}
\qquad
\begin{array}{r} 1098 \\ -\ 346 \\ \hline 752 \end{array}
$$

Closing the lesson [15 min]

Ask the children to explain their different results. Ask about the strategies that they used, and how they checked their solutions. Ask how many they found.

Assessment

Child's performance	Teacher action
Does not find a solution	Activities using subtraction patterns and problem solving
Finds one or two with help but generally unsure	Repeat Lesson 1, subtraction patterns using three-digit numbers
Finds a number of solutions confidently	The learning targets for this theme have been met

Homework

Collect some three-digit bus numbers and make up some subtraction problems. How many ways can you solve them? Do you use the same way when working mentally as with pencil and paper?

Decimals

Learning targets

On completion of this theme the children should be able to:

1 ➡➡ read and order decimal numbers
2 ➡➡ locate decimal numbers on the number line
3 ➡➡ add and subtract decimal numbers.

Before you start

Subject knowledge

Children commonly make two errors: decimal point ignored (DPI) and largest is smallest (LS). Pupils who make the DPI error treat decimal numbers as if they were whole numbers; typically they would treat 0.08, 0.26 and 0.1 as 8, 26 and 1. Children who make the LS error recognise that decimal numbers are different from whole numbers; they reason that the more decimal places a number has, the smaller it is, so 0.34 is smaller than 0.2. When the number of decimal places is the same, the error takes the form of the larger is smaller, thus 0.26 is smaller than 0.08. The key to understanding decimal numbers is place value. This alone makes sense of decimal notation, and any computation carried out with decimal numbers.

Previous knowledge required

The understanding of place value with whole numbers. The ability to add and subtract whole numbers.

Resources needed for Lesson 1

Sets of cards made up from Copymaster 66, place mat.

Resources needed for Lesson 2

Number lines marked in tenths from 0 to 10. Blank dice half of which are marked 0, 0.1, 0.2, 0.3, 0.4, 0.5 and the other half 0, 0.5, 0.6, 0.7, 0.8, 0.9, and a large place mat (enough for each child if possible). Dice marked 0–6 are also required.

Resources needed for Lesson 3

Copymaster 17.

Teaching the lessons

Lesson 1 ①

Key questions

Where is the best place to put 9 to make the biggest/smallest number?
How can you tell who has the biggest number?

Vocabulary

Decimals, tens, units, ones, tenths, column.

Introduction ⬚15min⬚

▓ Show the children a place mat, write in some digits, and ask the children the value of each of the digits. Write out another place mat using the same digits but in a different combination. Try some more numbers. Talk to the children about the value of a digit being determined by its place. Explain to the children that anything to the right of the decimal point is less than 1.

Activities ⬚30min⬚

👥 Make sure children have their own place mat. Working in pairs, tell the children to place the cards (from
▓ **Copymaster 66**) face down. Each child takes it in turn to place a card on their mat. They carry on taking cards until their mat is full, the object of the activity being to make the biggest number. Ask the children to record their numbers. Carry out the activity a number of times and then bring the class together. What numbers did they make? What strategies did they use? Talk about place value, 10s, units and tenths. The next part of the activity attempts to make the smallest number using the same method. Ask the children to record their results.

Closing the lesson ⬚15min⬚

▓ Ask the children to explain the results of the activity. Who found the biggest numbers? Who found the smallest numbers? Finish the session by writing numbers on a place mat and asking the value of the digits. Ask the children to read aloud to the class the numbers you have written.

Assessment

Child's performance	Teacher action
Cannot order decimals	Experiences using tenths, maybe on the number line
	Move on to next lesson
Can decide which number is bigger using tenths	
Can use decimals with confidence	Move on to next lesson

Lesson 2 ②

Key questions

How many jumps?
How many tenths?
How many tenths in 0.6, 0.8 … 1.7?

Vocabulary

Tenths, decimal place, whole number, counting on.

Introduction ⟦15min⟧

Show the children the number line. Talk about the gaps between the numbers and ask how they would fill them in. Remind the children of the previous lesson. Talk about decimal numbers and ask where they would place 0.5, 2.5 … on a number line.

Activities ⟦35 min⟧

Give each pair of children 3 dice (2 different decimal dice and a die marked 0–6), and a number line. Tell the children to take turns at throwing a decimal die and the ordinary die. Each player should use the 0–0.5 decimal die for their first go and the 0–0.9 die for their second go. Alternate for subsequent goes. They make a jump on the number line equal to the total of the two numbers. If 2 and 0.3 come up, they jump to 2.3. When it is their turn again and the dice show 3 and 0.6 they jump from 2.3 to 5.9. They should try to reach 10, and then begin at 0 again. Let them have several goes at reaching 10. The children should record their work.

Closing the lesson ⟦10min⟧

Ask the children to give examples of their games and show the number lines filled. Display your large place mat. Ask different children to fill in the columns using the numbers from their games. Make the connection between the number line and the place mat.

Assessment

Child's performance	Teacher action
Cannot complete number line	Repeat Lesson 1
Cannot count on accurately	More experiences using a number line
No difficulties	Move on to next lesson

Lesson 3 ③

Key questions

How many tenths?
Can you carry?

Vocabulary

Tenths, carry, make, units, column.

Introduction ⟦15min⟧

Review the previous lesson. Ask the children to suggest 2 decimal numbers to 1 decimal place. Carry out the addition, first adding the tenths column to try and make a whole, writing the remainder in the tenths column, carrying the whole to the units column. Do the same calculation using base 10 materials. Try a few more calculations. Stress the idea of making up the whole if possible and moving to the next column. Talk about grouping and counting in 10s. Every time you make a 10 it is moved 1 column to the left.

Activities ⟦30min⟧

Give every child a copy of **Copymaster 17**. Explain to the children that they have to complete the calculations on the copymaster in the same way they have just seen. Remind the children before they start to begin in the tenths column and try and make 10 tenths, then move to the column to the left. Support the children by asking the key questions.

Closing the lesson ⟦15min⟧

Ask the children to explain some of their results and how they carried out the calculation. Make sure the children share their ideas with each other. Finish the lesson by carrying out some of the calculations on the copymaster using base 10 materials. The more able child might be asked to subtract or find the difference between the numbers.

Assessment

Child's performance	Teacher action
Cannot add the decimals together	Repeat Lesson 1
Carries the remainder and writes down the 'ten'	Use base 10 materials to support, making sure the child physically carries out the addition
Carries out the addition correctly	The learning targets for this theme have been met

Homework

Help with the shopping, add up the till receipts, work out the change expected for given amounts of money. How many shopkeepers wrongly label their goods in your area? Carry out a small survey. Measure some of the furniture in your bedroom using metric measurements, e.g. 3.5m, 1.25m.

Addition and subtraction using money

Learning targets

On completion of this theme the children should be able to:

1➤➤use money and understands equivalent values
2➤➤use running records of expenditure
3➤➤add and subtract money using decimal notation.

Before you start

Subject knowledge

The addition and subtraction of money is restricted to whole numbers, if writing sums of money in pence, or to 2 decimal places, if the pound sign is used. Children need to recognise the current coinage and equivalent values.

Previous knowledge required

Has experience of addition and subtraction of whole numbers to at least 20. Can read, order, count whole numbers to 100. Recognises decimal notation and has used addition and subtraction with decimals. Recognises the coinage in current circulation.

Resources needed for Lesson 1

Plastic money (optional).

Resources needed for Lesson 2

Various catalogues, calculators (optional).

Resources needed for Lesson 3

Copymaster 18, counters or knights (chess pieces), labels marked with prices like £1.75, 175p (both correct) and £1.75p (incorrect).

Teaching the lessons

Lesson 1 ➀

Key questions

What combinations can you find?
How many different ways can you make 50p, 60p …?
Which 'price' gives the most/least combinations?

Vocabulary

Combination, money, total amount, coins, penny, cost, price.

Introduction | 10 min |

Ask the children how many different ways they can think of to make 20p, using 1, 2, 3, 4, 5 coins and so on. Record their results, and ask if they can see any patterns. If you exclude the use of 1p coins are there certain numbers of coins that don't appear when making 20p?

Activities | 35 min |

Tell the children they are going to find all the combinations of coins that can make £1, as with the 20p in the introduction, using 1, 2, 3, 4, 5, 6 coins and so on. After the children have been working for a while, bring the whole class together. Ask who has bought a drink from a drinks machine, or a bar of chocolate from a machine. Ask how much they paid. Explain that drinks machine manufacturers have a problem: there are a limited number of prices they can charge because people do not carry a vast number of pennies. Ask if they have ever seen a drinks machine that charged 33p, 47p, or 77p. Can they explain why?

Tell the children to find the prices they think that the machine manufacturers do charge. Hint at the number of coins and possible combinations chosen. You could carry out a survey on local machines with the children to verify their ideas.

Closing the lesson | 15 min |

Ask the children to explain any patterns they have found. Ask the children to justify their conclusions as to the 'best prices' from a drinks machine. End the session by asking the children, mentally, to work out possible combinations of coins for different amounts of money, e.g. 46p, 78p, 56p.

Assessment

Child's performance	Teacher action
Can find some patterns	More activities using money in different combinations
Can find combinations of coins, but has difficulty with ideas of 'best price'	Discuss the idea of 'best price' in the survey of machines
Makes a list of 'best prices' and can justify	Move on to next lesson

Lesson 2

Key questions

How much have you spent?
How much money have you got left?

Vocabulary

Million, thousand, hundred, running totals, most expensive, price, cost, value, amount, difference.

Introduction [15min]

Introduce the idea of a running total to the children. Read out a price list slowly and ask the children to mentally add the prices up. Do this a few more times with different amounts of money and totals. Tell the children they have £10 to spend and give them a list of objects they have 'bought' with the prices. Ask the children to find out how much money they have 'actually' spent and how much money they have left.

Activities [35 min]

Tell the class that they have £100 each to spend.
Show the children the catalogues and other price information, telling the children they must only buy goods from these sources. Explain to the children they must keep a running record of their expenditure and thus be able to tell you how much they have spent, and how much they have left to spend at any time. This task can be used with whole numbers only, depending on the resources you provide, or decimal numbers. The amount an individual has to spend may be varied, increasing totals to £500 or £1000.

Closing the lesson [10min]

Ask the children how much they have spent. Ask individuals to explain how they kept a running record. Ask what was the most expensive item and the least expensive. Ask the children to check each other's running record for accuracy, with a calculator.

Assessment

Child's performance	Teacher action
Cannot keep a running record	More experiences involving adding 3 or 4 numbers together
Can keep a running record but for a limited number of prices	Practice at keeping running totals in other contexts
No difficulties	Move on to next lesson

Lesson 3

Key questions

How would you add £0.01 … to £0.50 …?
What decimal fraction of a pound is 1 penny?
What decimal fraction of a pound is 10p?

Vocabulary

Decimal, whole, unity, money, tenths, hundredths, decimal point, sign, pound, pence.

Introduction [20min]

Show the children the prepared price labels. Ask the children if they can see what is wrong. Point out that the labels marked with both the pound and pence signs are incorrect – ask why. Explain to the children about the idea of unity. Point out it is only possible to write, e.g. £1.75 or 175p, not £1.75p. Which is the 'whole' – the pound or pence? Ask the children how they would write the following amounts of money using the pound sign: thirty-four pence, fifty pence, sixty-three pence and 1 penny. The solutions are £0.34, £0.50, £0.63 and £0.01. Remind the children about tenths and hundredths.

Activities [30min]

Show the children **Copymaster 18**. Tell the children they are going to play a game called the Robber Baron. Remind the children of how the knights move on the chessboard. Tell the children they have to take it in turns to be the robber baron. They move their knight 7 times in each turn. Each child collects the money on the square at the end of each full move of the knight. They have to collect the maximum amount of money by moving around the £1.00 square. They must total their collection at the end of each turn. Once a square has been visited it should be marked as it cannot be used again. The winner is the player with the most money.

Support the children by asking about the strategy they are using and how they calculate their moves.

Closing the lesson [10min]

Ask the children to explain the strategies they used to play the game. Ask for examples of the additions that have been carried out. Remind the children about decimal numbers and the correct way to write price labels.

Assessment

Child's performance	Teacher action
Confused by the notation	Needs experiences of using decimal notation
Has difficulty with the addition	Practice adding decimals
Understands both the notation and addition	Extend the activity to subtraction, or mentally

Homework

Keep a running total when out shopping. Survey the prices of a limited number of items from a number of shops, including the most expensive and the least expensive.

Number bases

Learning targets

On completion of this theme the children should be able to:

1➤➤ group and count in base numbers other than 10
2➤➤ add and subtract in base 6
3➤➤ add and subtract in base 8.

Before you start

Subject knowledge

The number system that is commonly used in our society is based on counting or grouping in 10; this is normally called a *denary system*. The counting of groups in 10 and multiples of 10s, gives us H T U. However, counting in 10s is not the only possibility. There are many ways of grouping or counting. In the past, children used to be given experiences of counting in different number bases because of the Imperial measuring system, e.g. 1 yard (3 feet), 1 foot (12 inches) and inches; and pre-decimal currency, e.g. 1 pound (20 shillings), 1 shilling (12 pence) and pence (d). As children, for the most part, do not encounter the above systems nowadays, base numbers should be explicitly taught in support of a greater understanding of place value.

Previous knowledge required

Experience of grouping and counting of objects. The ability to read, and count and order numbers. An understanding of addition and subtraction.

Resources needed for Lesson 1

Multi-link®, Copymaster 67.

Resources needed for Lesson 2

Egg boxes for 6 eggs, Copymaster 67, Multi-link®.

Resources needed for Lesson 3

Copymaster 19, multi-base materials (base 8) or Multi-link®.

Teaching the lessons

Lesson 1 ①

Key questions

How many groups of four can you make?
What happens after you reach 3?
What does 10 mean in alien numbers?

Vocabulary

Counting, grouping in four, base numbers, base four, groups of, how many?

Introduction 20min

▨ Ask the class to imagine that they are visiting another planet. The 'people' of this planet do not have 2 arms, they only have 1, and unlike us they do not have 5 digits (fingers and thumbs) on each hand but only 4 on their single hand. The person they are visiting is a farmer who farms 'quadladykes' a sort of sheep. She has to count them each day, but only counts in groups of four: 1, 2, 3, quad, quad + 1, quad + 2. Ask the children what number they think comes next. Discuss possibilities with the children. Share out some Multi-link® and tell the children each block represents a quadladyke. Can they tell you how many quadladykes there are in alien numbers? Re-arrange the number of Multi-link® and ask again – how many in alien numbers? Keep on doing this until the children understand the idea of grouping in 4.

Activities 30min

👥 Ask the children which numbers would exist on the alien planet (0, 1, 2, 3). Ask the children to justify their ideas. Give out the Multi-link® and tell the children it represents a flock of quadladykes. Ask them to tell you how many they have in alien numbers. Tell the children to mix up the Multi-link® with their partners and count again. Try this for a few more times. Ask the children to make a number square (4 × 4) to help them count the quadladykes. Copymaster 67 can be used for this as it gives space for alterations and corrections. Discuss before they start what it might look like and what numbers would exist. Ask the children to make up some alien sums for their partners to do.

Closing the lesson 10min

▨ Review the counting in 4s using the key questions. Ask the children to explain the sums they set as a challenge to their partner. Give them some sums to do mentally using base 4, e.g. 3 + 2 = 11 (base 4), 3 + 3 = 12 (base 4).

Assessment

Child's performance	Teacher action
Cannot follow the counting system	Activities that involve grouping and counting in base 4
Can follow the counting system but cannot calculate	More counting experiences
Can count and calculate in base 4	Move on to next lesson

Lesson 2

Key questions

How many groups of 6?
What numbers do you use in eggmatics?
What would an eggmatic number line look like?

Vocabulary

Base 6, counting in six, grouping in six, counting on, making six.

Introduction [20min]

 Remind the children about Lesson 1 and counting in 4s. Tell the children this time they are going to work in 'eggmatics'. Show the children the Multi-link®, and tell them it is not really Multi-link® but a brand new type of egg. Show the children the egg boxes and a pile of square eggs. Ask a child to sort the eggs into boxes and eggs. Point out that in the eggmatic world no number greater than 5 can exist, because after 5 eggs the sixth makes a box. Choose 2 children to be farmers. Give each child a pile of Multi-link® and egg boxes, asking them to sort/group their eggs. Then ask: how many? Ask the whole class how many boxes and eggs the farmers have altogether. Tell the children that six eggs = 1 box, 6 boxes = 1 carton, 6 cartons = 1 van load.

Activities [30min]

 Ask the children to do some eggmatic additions. After the children have completed the above task ask them to explain how they would subtract in eggmatics. For the more able child a multiplication square in base 6 might be attempted.

Closing the lesson [10min]

 Review the lesson using the key questions. Ask the children to calculate mentally in base 6, e.g. 5 + 4 = 13 (base 6), 5 + 5 = 14 (base 6).

Assessment

Child's performance	Teacher action
Has difficulty in counting in base 6	More experiences of grouping in base 6
Counts in base 6 but cannot calculate	More experiences of combining sets
Counts and calculates in base 6	Move on to next lesson

Lesson 3

Key questions

How many eights can you find?
How high can you count in spidermaths?

Vocabulary

Group in eights, count in eights, base 8.

Introduction [10min]

 Review and remind the children of the work they have already carried out in Lessons 1 and 2. Tell the children they are going to work on spidermaths. Ask what group or base they think they will be counting in. Use Multi-link®, or base 8 materials, and carry out some grouping activities. Ask which numbers are used in base 8 (0, 1, 2, 3, 4, 5, 6, 7). Point out that in spidermaths, 8 legs make 1 spider, 8 spiders make 1 web, and 8 webs make 1 cobweb. Ask the children what we use instead of spider, web and cobweb, in our counting system.

Activities [35 min]

Give out **Copymaster 19**. Explain the first part is simple spider additions. In the second part of the copymaster they have to use base 8 to create more calculations with the answer 15. They should be encouraged to use addition and subtraction. The more able child might try multiplication.

Closing the lesson [15min]

Ask the children to explain what they have found, giving examples of their work, especially from the last part of what they have done. Link the work with base 10. Remind the children that in everyday calculations we group in 10s and ask them why we do so.

Assessment

Child's performance	Teacher action
Has difficulty with base 8 addition	More sorting/grouping activities
Can use base 8 addition but not subtraction	Use a place mat, exchange activities
Can use both addition and subtraction	Try multiplication, other base numbers

Homework

Count the number of legs on the pet, dog or cat at home. Make up a base 4 number system for four-legged creatures. Draw an alien farmer.

Investigations

- Let the children add up the daily and weekly totals in the register. How does the total differ week by week?
- How much pocket money does the class get? Total all the pocket money in the class. Which class gets the most pocket money? What is the difference between the class that gets the most money and the one that gets the least?
- How many ways can you make your house number? Do the same using whole numbers, then decimal numbers. What is your house number in base 6, 8, 5, 4 ...?
- Collect some local 'take-away' menus and ask the children to cost their favourite meal. How much would all the meals cost? Who had the most expensive meal? And who the least expensive? What is the difference in price?
- A palindromic number is a number that is the same if you reverse the digits, e.g. 343. Choose a two-digit number, e.g. 34, and reverse the digits to make 43. Now add, 34 + 43 = 77. 77 is a palindromic number. Here is another example. Start with 48: 48 + 84 = 132. Now reverse 132 and add, 132 + 231 = 363, which is another palindromic number. Does this always happen? How many steps are there to a palindromic number?
- How much are you worth? Give each letter of the alphabet a price, starting with A = 1p and going on to Z = 26p. Work out the price of your name and your friend's name. Who has the most expensive name in the class?
- How many ways are there to get to 100? You must use a calculator, but can use each of the following keys once only in each go: 1, 2, 3, 4, 5, 6, 7, 8, 9, 0. But you can use + and/or − as many times as you like in a go. How close to 100 can you get? Keep a record of your results. If you find this one too easy, try making 999 using the same numbers (0–9 once only) using only addition.
- Pretend you are ill and the doctor has given you some pills. On the first day you take 1, on the second day 2, on the third day 3 and so on until on the tenth day you take 10. How many pills have you taken altogether? Can you find a quick way to work it out? What would be the number of pills if your treatment went on for 20 days, 50 days, 100 days?

Assessment

- Give the children addition squares to complete. Make jigsaws of the addition squares. Can they put them back together?
- Ask the children to make up as many number statements as they can, using addition and subtraction, that equal 10, 20, 100 or 100.
- Give the children 4 numbers. How many different calculations can they make using the numbers?
- Give the children some addition and subtraction problems they must do in their heads. Ask them to write down how they carried out the calculations.
- How many ways can they make 1 using tenths?
- Get the children to make up addition and subtraction games using a 100 square and 2 dice.
- Get the children to make function machines that 'undo' addition problems or subtraction problems.
- Explore the constant function on a calculator.

FRACTIONS

Fractions arose quite early in the history of number because of the frequency with which they were needed to solve real, everyday problems mainly to do with measurement. Even if you have a very small basic unit of length, for example, it is still the case that you will need to partition between whole numbers on the scale.

The fractions that our ancestors dealt with were generally of the kind $\frac{1}{2}, \frac{1}{4}, \frac{1}{3}, \frac{2}{3}$, and $\frac{3}{4}$. In fact these are still seen as the *common fractions* (along with others like $\frac{2}{5}, \frac{1}{6}$ and so on). In schoolbooks in the Victorian era and up until quite recently these common fractions were termed *vulgar*. Perhaps because fractions pose particular problems in both measurement and computation people have regularly given them special names. For example, where the numerator is smaller than the denominator we call it a *proper fraction*, and consequently when the numerator is larger than the denominator we call the fraction *improper*! And when we turn an improper fraction into whole numbers and a proper fraction we talk about a *mixed number*. Decimal fractions are centred on base 10 and $\frac{1}{10}, \frac{1}{100}$ and so on are the important fractions. All of this can be very confusing for children.

In teaching and learning about fractions we need to really focus on the ways in which fractions are used. There are three main ways.

1 Fractions are used to represent parts of a whole, or parts of a set. So we might have $\frac{1}{4}$ of the cake, or $\frac{1}{4}$ of the sweets in the bag.

2 Fractions are used as a way of writing division problems. So if we want to model 192 seats arranged in 12 rows we might write $\frac{192}{12}$.

3 Fractions appear in discussions of ratio. So if the scale for a model is 1:3 we often see this depicted as a $\frac{1}{3}$ scale model. Fractions and ratios are clearly linked; indeed the Ancient Greeks saw fractions like $\frac{3}{5}$ as being in the ratio 3 to 5.

All these ways of using the fractional form have to be encountered, discussed and understood by the children. If they are not then confusion can easily arise, particularly in relation to division where there are a number of setting-out conventions that the children will encounter. It is worth taking time on this, as the use of a division form is fundamentally important when children get on to certain aspects of algebra and data handling.

Other important ideas in work with fractions are *equivalent fractions* and the language used when trying to explore the computation of fractions. Without a good grasp of equivalent fractions the children will not be able to manipulate fractions as they progress. Associated with this, of course, is the reduction of fractions to the lowest expressible value ($\frac{2}{4} = \frac{1}{2}$) and the lowest common denominator when computing fractions.

Whilst many primary children will not progress beyond the addition and subtraction of fractions it is important for those who do move on that the words used are appropriate. In the multiplication of whole numbers we can, for example, talk of 3 lots of 2 for 3×2. Similarly, we can talk about $8 \times \frac{1}{4}$ as being 8 lots of $\frac{1}{4}$ which is 2.

Halves and quarters

Learning targets

On completion of this theme the children should be able to:

1 ➤ identify and produce halves and quarters in a variety of ways

2 ➤ use halves and quarters written symbolically

3 ➤ double and halve numbers.

Before you start

Subject knowledge

There are two key elements that the children need to develop in early work on fractions. Firstly, fractions such as a half and a quarter are commonly used in describing the division of one whole thing (cake, pizza, apple) into two or four equal parts. This raises major questions about the importance of the term 'equal'. We do hear people talk about the 'biggest half' and it is important that children should appreciate that this is nonsensical in mathematical terms. The other important element in the children's understanding is the use of fractions in relation to a set of objects: e.g. from a box of 12 pencils we can take half, i.e. 6 pencils (or, indeed, a third, a quarter or a sixth).

Previous knowledge required

A skill in using rulers, scissors and in folding paper. Some knowledge of times tables particularly the 2× table.

Resources needed for Lesson 1

Rulers, coloured pencils, strips of plain paper 10cm wide, Copymaster 20.

Resources needed for Lesson 2

Scissors, rulers, Copymaster 20.

Resources needed for Lesson 3

Copymaster 21.

Teaching the lessons

Lesson 1 ①

Key questions

Where do we use halves and quarters?
How many ways can you make a half and a quarter of a given object?
What is the relationship between a half and a quarter?

Vocabulary

Half, halves, quarter, quarters, fraction, square.

Introduction 20min

Ask the children where they have heard, or used, the term 'half'. Make a list on the board. Get them to try to group these, e.g. in relation to telling the time, sharing out things, and dividing into parts. Emphasise the fact that halves and quarters are the most common fractions that we use in everyday life. Explain to the children that it is possible to 'divide' a whole shape into halves and quarters in a variety of ways.

Activities 30min

Give the children **Copymaster 20**. The challenge is for them to produce 2 different ways of dividing a square in half, and in quarters. They can use rulers and coloured pencils. Bring the children back together and get individuals to share their solutions with the rest of the class. How many different ways have the class found? Give out 4 strips of paper to each of the children and ask them to find 2 ways of folding the paper strips in half and in quarters. Bring the class back together to share solutions.

Closing the lesson 10min

Review what should have been learned, with the children demonstrating their understanding through your use of the key questions.

Assessment

Child's performance	Teacher action
Skill problems in use of ruler and/or accurate folding	Give practice at drawing lines with a ruler and offer strategies for accurate folding
Understands halves and quarters in this context	Move on to next lesson
Clearly understands halves and quarters and talks about connections between them	Move on to next lesson

Lesson 2 ②

Key questions

How do you write a fraction using numbers?
How many ways can you make three-quarters using halves, wholes (units) and quarters?
Do the halves or quarters need to be exactly equal?
Are two quarters equal to one half?

Vocabulary

Three-quarters, equal.

Introduction | 15 min |

 Remind the children of how much they already know about the fractions a half and a quarter. On the board draw a square and with the children's assistance divide it into two halves. Write 'half' and 'two halves' by the diagram. Explain that we can think of the square as being one whole and that we have divided the whole into 2 halves. Write: 'One whole divided into two halves'. Explain that we could write this as one whole divided by two, or 1 divided by 2: we can write this as $\frac{1}{2}$. Now get the children to take you through the same process for arriving at $\frac{1}{4}$.

Activities | 30 min |

Give out Copymaster 20. Invite the children to colour in halves and quarters and to write down the words and symbols associated with $\frac{1}{2}$ and $\frac{1}{4}$. Give the children another copy of Copymaster 20. Now they need to cut out the fractional parts and use them to make as many whole squares as they can.

Closing the lesson | 15 min |

Get some of the ways in which the children have made a whole square and put these on the board in words, e.g. 'a half and a quarter and a quarter', then write these out in symbols using the words as you go, e.g. $\frac{1}{2} + \frac{1}{4} + \frac{1}{4}$. Ask the children to go through speaking the words that match the symbols and signs with you.

Assessment

Child's performance	Teacher action
Mixes the wrong symbols with the words 'half' and 'quarter'	More practice in shading, cutting and folding shapes, and discussion
Completes the tasks on words, symbols, and fractions to make the whole	Move on to next lesson
Uses solutions such as $1 - \frac{1}{4}$ or $1 - \frac{1}{2} + \frac{1}{4}$ to make $\frac{3}{4}$.	Move on to next lesson but prepare some additional challenges that combine operations in the context of doubling and halving

Lesson 3 ③

Key questions

What is half of ...?
What is double ...?
What is twice as much?
How did you work out your answer?

Vocabulary

Halving, doubling, multiply, divide.

Introduction | 20 min |

 Ask the children some mental arithmetic questions of the following type: What is double 2? What is twice 3? What is a half of 10? Use numbers that produce answers up to 20. Write down these number facts on the board and go over them with the children.

Half of

12 is 6	8 is 4	14 is 7.
6 is 3	10 is 5	20 is 10.

Double

5 is 10	7 is 14	2 is 4.

When you feel that the children are fully attuned to the doubling and halving idea move on.

Activities | 20 min |

Give out **Copymaster 21**. Get the children to work through the problems, connecting their answers with arrows as in the example.

Closing the lesson | 15 min |

Get the children together and go over the answers with them using individual contributions.

Assessment

Child's performance	Teacher action
Makes errors in doubling and/or halving	Practice needed in 2× table
Completes all tasks correctly	Try some of the investigations/games and the assessment ideas
Indicates that can go beyond quantities given	Offer some of the investigations and extension materials from available texts

Homework

Conduct a survey asking friends and family where they use the terms 'half' and 'quarter'. Find out whether some people, because of their jobs or interests, are more familiar with particular uses of fractions. Practice the 2×, 5× and 10× tables.

Developing fractions

Learning targets

On completion of this theme the children should be able to:

1 ➤➤ apply knowledge of halves and quarters to a range of shapes
2 ➤➤ discuss important features of simple fractions
3 ➤➤ apply what they know about fractions to extend their understanding of different types of fractions.

Before you start

Subject knowledge

There are special terms used in describing the parts of a fraction when written symbolically, and particular terms for types of fractions. We use the word 'numerator' for the quantity that is above the line and 'denominator' for the quantity below. Common fractions such as:

$$\frac{1}{2} \quad \frac{1}{4} \quad \frac{2}{3} \quad \frac{5}{8} \quad \frac{3}{4}$$

used to be called 'vulgar' fractions and this term is still in use in some books. Such common fractions are also examples of 'proper' fractions in which the denominator is larger than the numerator. 'Improper' fractions such as $\frac{5}{3}$ have a larger numerator than denominator. Improper fractions can be written as numbers of wholes and fractions, e.g. $\frac{5}{3}$ can be written as $1\frac{2}{3}$, and would then be termed 'mixed'.

Previous knowledge required

Understanding of halves and quarters and their symbols. Has worked on combining halves and quarters to make a whole, and three-quarters.

Resources needed for Lesson 1

Mirrors, rulers, Copymaster 22.

Resources needed for Lesson 2

Copymaster 23, Multi-link® or similar construction apparatus.

Resources needed for Lesson 3

Sets of objects, the number of which can be used to make simple fractions, counters, Copymaster 24.

Teaching the lessons

Lesson 1 ①

Key questions

How can you divide these shapes into halves and quarters?
What other common fractions do you know?
What is the connection between a whole, $\frac{1}{2}$, and $\frac{1}{4}$?

Vocabulary

Appropriate use of fraction names.

Introduction [25min]

▓ Remind the children of work they have done previously on halves and quarters of squares. Draw a kite on the board and invite suggestions as to how this might be divided into halves. Remind the children about equal fractions. Ask the children whether the kite can be divided into quarters? Repeat this exercise for a rectangle. Tell the children that shapes may sometimes only be divided into two equal halves.

Activities [25min]

▓ Give out **Copymaster 22**. The children have to see in what ways they can divide each shape into equal fractions – halves or quarters or both. The results may look like this:

Closing the lesson [10min]

▓ Get the children to share their ideas and solutions. This activity should provoke a lot of mathematical effort and thinking. While closing the lesson be encouraging about the process the children have gone through.

Assessment

Child's performance	Teacher action
Struggles to divide any of the shapes into halves or quarters	Undertake appropriate parts of the work on halves and quarters in this section
Comes up with at least one appropriate fraction in each case	Move on to next lesson
Develops a range of alternative ways of dividing up the different shapes into equal fractions	Move on to next lesson or to more work on symmetry; consider some additional challenges from other texts

Lesson 2 ②

Key questions

What fraction of the whole is this?
What connections can you see between some of the fractions you know?

Vocabulary

Half, quarters, equal; eighths, tenths, fifths and others the children suggest.

Introduction 20min

 Remind the children of the work they have done on wholes, halves and quarters. Tell them that there are more ways that we can divide a whole into equal fractions. Using a strip of paper show the children a half by folding, then a quarter and then an eighth. Draw the fold effects on the board so that all the children can see what has happened. Write up the fractions $\frac{1}{2}$, $\frac{1}{4}$, then $\frac{1}{8}$ as a reminder of how we write fractions.

Activities 25min

 Give out **Copymaster 23**. Get the children to follow the instructions to colour in fractions of each strip. The Multi-link® or similar apparatus should allow the children to model the exercise in 3 dimensions if this is appropriate.

Closing the lesson 15min

 Bring the children back together and then use volunteers to give their answers so that the children can carry out a self-checking process. Go over the challenges on $\frac{3}{10}$, $\frac{2}{5}$ and $\frac{5}{8}$ particularly carefully. See whether the children can make connections between not only $\frac{1}{2}$ and $\frac{1}{4}$ but also $\frac{1}{5}$ and $\frac{1}{10}$, and $\frac{1}{3}$ and $\frac{1}{6}$.

Assessment

Child's performance	Teacher action
Cannot complete more than the most simple of the challenges	Give additional paper-folding activities then retry the exercise
Tackles all of the tasks offered	Move on to next lesson
Comes up with own extension ideas, or makes connections beyond those required	Get child to explain own ideas to friends then move on to next lesson

Lesson 3 ③

Key questions

How can you obtain fractions of sets of objects?
What is a mixed fraction?

Vocabulary

Fraction names, mixed fraction, proper, common, improper, vulgar (depending on the scheme in use).

Introduction 15min

The work the children have done on fractions has concentrated on their understanding of fractions of a whole. Here the children are given the opportunity to see that fractions of sets of objects are important too. Use the collections of objects that you have assembled to identify 'how many?' Use $\frac{1}{2}$, $\frac{1}{4}$, $\frac{1}{3}$, $\frac{1}{5}$ as appropriate to the number of objects in the sets you choose. Use the opportunity to ask questions in which the numerator is greater than 1, e.g. $\frac{2}{3}$ and $\frac{3}{5}$. Also ask the children about $\frac{2}{4}$, $\frac{3}{6}$ and so on. Give the children **Copymaster 24**. Take them through the first challenge as a group.

Activities 25min

The children can then go on and try the other numbers suggested on Copymaster 24. Encourage them to use the counters to explore factors then to try to find $\frac{1}{2}$, $\frac{1}{4}$, $\frac{1}{3}$, telling them that in some cases it is not possible to divide the counters into equal piles.

Closing the lesson 20min

Ask questions that move into mixed fractions. For example, how many sets of pencils do I need for a class of 35 children if there are 10 in a set of pencils? If it is 100m around part of our playground then how many times would I need to walk around it to go 350m? If I had 3 bars of chocolate to share between my 2 nephews how many bars would they each get? See if any of the children can make up these sorts of questions themselves.

Assessment

Child's performance	Teacher action
Completes the exercise on sixths with help	Move on to next chosen theme but give further practice on similar tasks
Completes all major parts of the copymaster	The learning targets for this theme have been met
Produces a wide range of appropriate ideas	Offer extension opportunities

Homework

Find a recipe that tells you how many people it is intended for. Write out the recipe for twice as many people. Using 4 identical rectangles how could they be cut up and shared equally between 7 people?

THEME 16 | Equivalent fractions

Learning targets

On completion of this theme the children should be able to:

1 ➤ locate and find fractions on the number line
2 ➤ work out some fraction families
3 ➤ find equivalent fractions by at least one method.

Before you start

Subject knowledge

A key idea in the use and manipulation of fractions is that of equivalence, central to our capacity to understand how to carry out computations involving fractions. To fully understand equivalence there is a range of related topics to address. The 2 common methods for finding equivalent fractions make use of all of these.

The first method involves using the counting numbers for the numerator and multiplying the denominator by the same factor, 2, each time e.g.: $\frac{1}{2}$ $\frac{2}{4}$ $\frac{3}{6}$ $\frac{4}{8}$ $\frac{5}{10}$

The second method involves multiplying both numerator and denominator by the same factor. So, for example, multiplying each by 3 we get: $\frac{1}{2}$ $\frac{3}{6}$ $\frac{9}{18}$ $\frac{27}{54}$ $\frac{81}{162}$

The *reduction* of fractions is the inverse of these methods. For example, if we have a result of a computation such as $\frac{25}{35}$ then we can divide numerator and denominator by 5 to achieve $\frac{5}{7}$.

Previous knowledge required

Knowledge of common fractions. Understanding of fractions as equal parts of the whole. Use of the number line.

Resources needed for Lesson 1

Copymaster 25.

Resources needed for Lesson 2

An envelope for each group containing card strips as follows – one left whole, one cut into halves, one cut into quarters, one cut into eighths and one cut into tenths – rough paper, Copymaster 60.

Resources needed for Lesson 3

Flash cards with different common fractions written on them, made up from Copymaster 26 by cutting out the fractions and sticking them on thin card.

Teaching the lessons

Lesson 1 ①

Key questions

Where on the number line is …?
What do I get if I add (subtract) this quantity?

Vocabulary

Appropriate fraction names, 'mixed' fractions.

Introduction | 20 min

Start by asking the children about the number line. What can they tell you about number lines? Draw an example on the board and, with the class, do some addition and subtraction of whole natural numbers. Introduce them to the idea that between each natural number there is a continuous line and we can label parts of it. Using the number line determine with the children that mid-way between, say 2, and 3, must be the half-way point which we call $2\frac{1}{2}$. Repeat this for other common fractions such as $\frac{1}{4}$ and $\frac{1}{10}$ making appropriate diagrams as you go.

Activities | 25 min

Give out **Copymaster 25** and invite the children to tackle the questions. Go around the class giving support and encouragement. As appropriate, remind the children of common fractions which have even denominators, and those that have odd denominators. This may help them to move flexibly from one challenge to another.

Closing the lesson | 15 min

Choose a selection from the given number line questions and go over them. Ask for individual suggestions for other number line parts and find out how the children have divided these into equal fractions.

Assessment

Child's performance	Teacher action
Cannot determine fractional parts consistently or with any confidence	Do more work on number lines using natural numbers then using halves, then quarters

| Completes the given tasks but shows uncertainty about some of the open-ended questions | Give more practice |
| Completes all parts of the task accurately and with clear insight | Move on to next lesson |

| Completes the task with at least 2 fraction families noted | Move on to next lesson but observe how the child sustains understanding of the work |
| Completes the task and goes on to partition squares | Move on to next lesson |

Lesson 2 ②

Key questions

What sorts of numbers can be shared into equal halves? (Thirds? Fifths?)
How do we write a half (a third, a fifth) in sixths (ninths, tenths)?

Vocabulary

Fraction names, equivalent, fraction family.

Introduction ⏱ 15min

Start by asking what the children know about different fractions, e.g. those with odd-numbered denominators and those with even. Also encourage them to talk about mixed fractions and fractions like $\frac{1}{2}$ and $\frac{3}{6}$ which mean the same amount. Indicate that in this lesson the children are going to extend their understanding of fractions which are equivalent to each other. Show them, on the board, an example of a fraction family like this one.

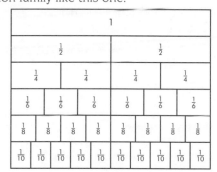

Activities ⏱ 30min

Give out pre-prepared strips of card in envelopes as suggested in the resources. The children should build up as many patterns as they can using the available strips. They should record these patterns in note form on rough paper for later transcription. When the pairs are ready give out copies of **Copymaster 60** and invite them to draw squares, partition them and write in the fractions that each partition represents.

Closing the lesson ⏱ 15min

Reinforce the introduction through the use of the children's efforts. Emphasise the idea of equivalence, and indicate some of the factors that make certain fractions part, or not part, of a particular family.

Assessment

Child's performance	Teacher action
Cannot develop more than one fraction family from the strips	Move to more practical activities geared to developing basic idea to do with what fractions

Lesson 3 ③

Key questions

How can we generate a set of equivalent fractions?
How many ways can you depict a given fraction?
What is the result of adding, e.g. $\frac{1}{2} + \frac{1}{4} + \frac{1}{4}$ or $\frac{2}{8} + \frac{3}{8} + \frac{3}{8}$?

Vocabulary

Equivalent fractions, equivalence.

Introduction ⏱ 20min

Review the idea of equivalent fractions. Then put the following fractions on the board in a random order: $\frac{1}{2}$, $\frac{3}{6}$, $\frac{1}{10}$, $\frac{10}{100}$, $\frac{1}{5}$, and $\frac{2}{10}$. Get the children to pair them, discussing all the responses, right or wrong, in order to reinforce the rules for equivalence. Next show how equivalent fractions can be developed on the basis of multiples using the natural numbers as the numerator in the first instance (see *Subject knowledge* above for examples). Then show the second way in which equivalents might be generated.

Activities ⏱ 25min

Give the children fractions on card made up from **Copymaster 26**. The children should work out equivalents of the fractions they have, using one or both of the methods you have outlined.

Closing the lesson ⏱ 15min

Test the knowledge the children have by quizzing them on the work they have done.

Assessment

Child's performance	Teacher action
Needs a lot of help to complete part of one set of equivalent fractions	Give more practice on fraction families using ideas from previous lessons/units
Completes the work using one of the two possible methods shown	Record the need to return to similar work in the future
Produces accurate examples of both ways of generating equivalent fractions	The learning targets for this theme have been met

Homework

Construct other fraction families. Develop the numerator/denominator lines further. Design and construct a set of questions on equivalent fractions.

THEME 17 | Fractions, decimals and percentages

Learning targets

On completion of this theme the children should be able to:
1 ➡➡ explain and use decimal fractions
2 ➡➡ interpret simple percentages
3 ➡➡ link decimal fractions and percentages.

Before you start

Subject knowledge

In working across and through ideas involving fractions, decimal fractions and percentages we are dealing with a lot of closely related ideas. These include equivalent fractions, work in the manipulation of decimal numbers, and the ratio that is termed *percentage*. Whilst there are a number of methods for conversion it is more important that children can work practically to see relationships rather than memorising a particular way of 'doing percentages'. The key ideas are that:

- percentages are ratios based on hundredths
- decimal fractions and percentages have a clear relationship because of their roots in base 10, and
- some common fractions can readily be expressed in percentages and decimal fractions (although many cannot, e.g. $\frac{1}{3}$, $\frac{1}{7}$).

The children also need to learn that we can convert fractions that are not decimal fractions into percentages. For example, what percentage is 4 of 16? Here we have to work through from $\frac{4}{16}$ to $\frac{1}{4}$ to 25%.

Previous knowledge required

Equivalent fractions.

Resources needed for Lesson 1

Copymaster 68, an overhead transparency of a 100 square or a drawing on the board.

Resources needed for Lesson 2

Copymaster 68, Copymaster 27.

Resources needed for Lesson 3

No special resources required.

Teaching the lessons

Lesson 1　①

Key questions

What does 'percentage' mean?
What does, for example, 25% mean?
What is one connection you can see between fractions and percentages?

Vocabulary

Percent, percentage, fraction names.

Introduction　20min

▦ Using a blank 100 square, talk about the fractions you can produce by colouring or crossing out numbers of the squares. For example, colour 15 squares and point out the fraction 'fifteen hundredths', then 10 more for '25 hundredths', and so on. Some children may observe that these fractions could be reduced, e.g. some will see readily

that $\frac{50}{100}$ is a half of the squares. However, the important idea here is to be looking for common denominators – in this case hundredths.

When you have a few examples of fractions of the 100 square then draw the children's attention to this common denominator. Explain that there is a way of describing these particular fractions by the use of the term 'percentage'. Give examples in fractional and percentage form: $\frac{50}{100} = 50\%$, $\frac{15}{100} = 15\%$, and so on. Explain that we can think of percent (or per cent) as meaning 'out of a hundred'. Hence 15% means 15 out of a 100.

Activities　15min

▦▦ Give copies of **Copymaster 68** to the children together with 100 counters or cubes. Get them to try the same exercise as you have been doing. They can decide on their own fractions and write these down together with the percentage equivalents. They should try at least 5 examples.

Closing the lesson `25 min`

 Get pairs to offer one of the percentages they have identified. Use these to reinforce your introduction. Then give the children some examples of where we can use percentages. These should include examples like:

- 100 people went to a concert, 40 of them really enjoyed it but the others thought it was not very good. What percentage liked it? What percentage did not?
- In assembly one morning there were 100 children. 25 of them were 8 years old, 30 were 9 and the rest were 10. What percentage were 10-year-olds? What percentage of the whole group were all of the 8 and 9-year-olds?

Assessment

Child's performance	Teacher action
Can identify fractions and percentages but finds it hard to develop own examples	More practice with hundredths using the 100 square; if this does not work then may have to return to earlier fraction work
Follows the lesson with confidence	Move on to next lesson
Follows the lesson with confidence and is clearly able to respond creatively to the examples in the closing of the lesson	Move on to next lesson and consider giving some more examples for child to work on independently

Lesson 2 ②

Key questions

What is, for example, $\frac{1}{2}$ as a decimal fraction?
What is, for example, 75% as a decimal fraction?
What is a given fraction in decimals and percentages?

Vocabulary

Common fraction names, percent and percentage, decimal fraction.

Introduction `20 min`

 Using a 100 square and a number line revise what the children should know about equivalent fractions. Ask the children to indicate where $\frac{1}{2}$ is and make suggestions about equivalent fractions using tenths and hundredths. Do the same for $\frac{1}{5}$ and $\frac{3}{10}$. Explain that we can use $\frac{5}{10}$ or $\frac{50}{100}$ or $\frac{1}{2}$ (all equivalent) for 50% of the space between 0 and 1 on our number line. This amount is said to be 0.5 (half way along the line). See if they can determine where $\frac{1}{4}$ might be and then see whether they can develop a decimal fraction for this. They should know that $\frac{1}{4}$ is equivalent to $\frac{25}{100}$ and this could be called 25%.

Activities `30 min`

 Give out **Copymaster 27**. Get the children to fill in the missing quantities.

Closing the lesson `10 min`

 Sample from the sheets checking some of the missing numbers. Use this as an opportunity to revisit important ideas that were in the introduction to this section.

Assessment

Child's performance	Teacher action
Can complete first tables but finds later ones too demanding	Can move on but these or similar tables to be presented again at a later date
Completes all tables with only minor errors	Move on to next lesson but check on errors
Completes all tables without errors	Move on to next lesson

Lesson 3 ③

Key questions

What is the percentage?

Vocabulary

Percentage, fraction, decimal fractions.

Introduction `25 min`

 Explain to the children that in this lesson they are going to learn how to calculate percentages for groups where there are not 100 members. Use the following in the order given:

- $\frac{1}{4}$ of a class of 24 children have a cold. What percentage of children is that?
- 10% of the parents at a school meeting wanted to change the uniform design. There were 200 parents at the meeting. How many parents wanted a change?
- 50 children stay for packed lunch one day. Of these 50 there were 10 who had cheese sandwiches. What percentage of children had cheese sandwiches?

Activities `25 min`

 Ask the children to make up some similar problems for themselves.

Closing the lesson `10 min`

 Get the children to talk through some of their ideas. Collect in their ideas to be collated for a class booklet.

Assessment

Child's performance	Teacher action
Responses to be evaluated as the booklet is produced	Devise a set of differentiated activities based on the problems and produce a class booklet

Homework

Ask the children to produce an exhaustive story of one or more fractions. How many ways can you write them? Where might you meet them? Interview family members about the places they meet percentages. How do they make decisions about bargains? For example, is 20% off the marked price better than 'buy 2, get 1 free'?

Investigations

- Using pairs of blank dice write common fractions and decimal fractions on the faces using a washable pen. The children have to roll a pair of dice and sum the quantities rolled, e.g. $0.5 + \frac{1}{2}$, and $\frac{1}{4} + 0.75$.
- With the help of parents and colleagues find out how far those children who get to school by car travel. This should be in decimals – the car's odometer can be used to get the figures. Using a map of an appropriate scale, draw in circles of different diameters, which will allow discussion about distance 'as the crow flies' and as the roads take us. Use this discussion to work out some ideas using averages.
- Get the children to produce a range of measures of length as accurately as they possibly can. How many decimal points can they work to?
- Make a display of animals and plants from around the world. The choices should be based on particular attributes relating to physical size. So we need largest and smallest snakes, spiders, crabs, elephants, and so on. Then ask the children to do some comparisons of the dimensions in fractions, decimals and percentages as appropriate.
- Use the attendance registers in school to determine and explore a range of questions that can be answered using fractions and percentages. For example, are there times of the year when more people are absent? What is the average attendance of the children? What percentage is within 5 days of this average? If it is appropriate, some good work can also be done on the dinner registers – favourite days, weeks? What about staying for dinner and age? And so on.
- Make a collection, with the help of the children, of special offers and sales that might be available in your area. We all now receive a lot of mail advertising money off, and/or reductions. These make a good start. The children can work out costs, discuss bargains and wrestle with how one can make choices. They should also tackle problems such as: The cost of a pair of shoes is £30 and a sweater is £15. There is a sale in which you get 10% off every item. Does it make any difference whether we buy both together and have the 10% off the total or whether we buy them separately?
- Holiday brochures are also a good source of ideas about discounts, and working in groups, the children could have a go at working out the cost of a holiday comparing different offers.
- Get the children to develop ideas for surveying others about preferences. For example:
  - TV programmes
  - food and drink
  - local developments such as a new supermarket
  - best times for learning mathematics, etc.
  Use the data collected to look at percentages.
- Make a set of dominoes – the children could help – on the ends of which there is a fraction, a decimal or a percentage. Play dominoes matching the ends. This activity could be linked to an investigation of conventional dominoes. How many are there in a set? How and why are they arranged in the way they are? How many do we need for our dominoes? What percentage of the set has a 'six' on them?
- The children can make conversion tables that allow you to read off the fraction/percentage, decimal fraction/percentage, or fraction/decimal fraction.

Assessment

- Give the children a set of numbers (orally or written on the board). They have to produce either the double or the half (or both) of each of the given numbers.
- Repeat with quarters and quadruples. This assesses their $4\times$ table and their understanding of the relationship between multiplication and division.
- Undertake frequent oral assessments involving mental calculation. Topics should include: greater and smaller (e.g. which is greater 0.5 or 0.05), simple additions of combinations of $\frac{1}{2}$, $\frac{1}{4}$, $\frac{3}{4}$, and mixed numbers which include those fractions, e.g. $1\frac{1}{2} + \frac{1}{2}$, $1\frac{3}{4} + \frac{1}{4}$, $2\frac{1}{4} + \frac{1}{2} + \frac{1}{4} + \frac{1}{2}$, $\frac{3}{4} + \frac{1}{4}$ and so on.

- Test on appropriate multiplication tables in support of children looking at equivalent fractions.
- Get the children to write out the inverse of given multiplications and divisions, e.g. $1 \div 2 = \frac{1}{2}$ and $\frac{1}{2} \times 2 = 1$.
- Give problems involving money and percentages.
- Get the children to divide given shapes into equal fractions.
- Make up fraction families using rods, blocks or other apparatus that might be available.

MULTIPLICATION AND DIVISION

The processes of multiplication and division build on earlier work on addition and subtraction.

Multiplication can be seen as repeated addition in many circumstances so the children can be encouraged to see, for example, that 3×2 has the same effect as $2 + 2 + 2$. We can use appropriate language to support this perception saying, for example, '3 lots of 2'. We could also, of course, state that there are '2 lots of 3' and this demonstrates one of the key ideas that the children need to develop here – that multiplication is *commutative* so $3 \times 2 = 2 \times 3$ is a true statement.

The use of the idea of repeated addition is not one, though, that readily extends into all situations where we are multiplying. For example, 8×0.2 can be explained as the repeated addition of eight 0.2s to give a total of 1.6 but it helps to encourage children to see that multiplying by a decimal number brings about a *reduction*. This will help when they get to problems like 0.4×0.3 where the answer of 0.12 seems inconsistent with addition. So it is the case that the multiplication of whole numbers should not only be viewed as repeated addition but also as *enlargement*. The scaling effect of multiplication needs to be given due attention as the children progress, and the language of scale and ratio should be deployed as well as that of addition.

Language is equally important when we come to division and in the early stages the idea of 'shares' is one that can be usefully encouraged. Children know a lot about sharing and their experience will support division problems that are phrased like: There are 8 apples to be shared between 4 children. In symbolising this as $8 \div 4$ the children need to get the order right because $8 \div 4 = 4 \div 8$ is not a true statement; division is *not* commutative. As with multiplication, division also needs to be expressed in ways other than sharing. Division can be seen as the inverse of multiplication so that $15 \div 3 = 5$ can be seen as leading to the statement that $5 \times 3 = 15$. As we know that multiplication is commutative it makes sense then to appreciate that we could divide our 15 into 3 groups of 5, or 5 groups of 3. Grouping is not the same idea as sharing. Division also relates to ideas of ratio and scaling as with multiplication.

There are other laws or rules, besides commutativity, that children need to appreciate when handling multiplication and division. Particularly we are concerned with the *associative* and the *distributive* laws. Associativity will have been encountered in work on addition. For example, in solving $8 + 9 + 2$ it does not matter whether we do $(8 + 9) + 2$ or $8 + (9 + 2)$. The same applies with multiplication, so $(3 \times 2) \times 4 = 3 \times (2 \times 4)$. In many situations we will choose to associate numbers which simplify what we have to do. Similarly, we can often simplify matters by using the distributive law. So that in considering a problem like 13×43 we can redistribute it as $(10 \times 43) + (3 \times 43)$. All these processes are important, and there are links between them.

In presenting these laws, use has been made of brackets and children do need to know the conventions that apply to the use of brackets. The part in brackets is done first so that whilst $(4 \times 3) - 2 = 10$, $4 \times (3 - 2) = 4$. However, in creating their own calculations or having to solve a problem presented without brackets the children also need to know that there are precedence rules. So faced with $5 \times 6 - 4$ we would do the multiplication first so $(5 \times 6) - 4$.

Finally, although there are lessons within this section which are intended to support the teaching of a range of algorithms, we would urge that such algorithms not be treated as the only 'right' way to do multiplication or division. There is clear evidence that a constraining use of one algorithm can reduce the effectiveness of children in problem solving.

Multiplication tables

Learning targets

On completion of this theme the children should be able to:
1 ➡ give accurate answers to questions about multiplication tables
2 ➡ use the commutative nature of multiplication
3 ➡ use their knowledge of multiplication bonds to solve numerical problems.

Before you start

Subject knowledge

There is no doubt that memorising the common multiplication tables is of paramount importance in developing the ability to handle numerical problems. There is an understandable reluctance to see children having to learn their tables by rote although this was the common approach some years ago. We want the children not only to be able to 'chant' but also to understand. To this end the teaching of tables should be planned and must contain a variety of learning opportunities for the children. We have to provide a range of activities, regularly and over time, in order to increase the probability that the children will really know their tables. Unlike the other themes in this section there are not three lessons: rather we have listed a variety of teaching ploys. We feel that if this full range of possible approaches is used throughout the early period of the Key Stage then more children will experience greater success.

Previous knowledge required

Must have worked on addition and place value, and be experienced in counting and tallying.

Resources needed

Counters, squared paper (Copymaster 67), number lines, Copymaster 69, Copymaster 28.

Teaching the lessons

Key questions

What is times …?
Which multiplications have the same product?
What is the pattern?
How quickly can you count on in, say, 4s?

Vocabulary

Multiplication, times tables, product, counting on.

Activities |5min| |10min| |15min|

Times vary; a whole lesson could be devoted to patterns.

- Using counters and squared paper, get the children to make rectangles. For example 12 counters could make
  1×12, 2×6, 3×4, 4×3, 6×2, and 12×1 rectangles. Not only will this help with tables but it also supports an understanding of commutativity.

- Counting on in 2s, 3s, 4s, 5s and 10s should be practised. Such work clearly supports the accurate solving of addition problems mentally. Number lines can be used very profitably in this case.

- Explore the multiplication of a number by itself. The product is, of course, a square number. The square numbers should be recognised and known in their own right.

- There is nothing wrong with the chanting of tables as part of the memorisation process. However, chanting needs to be linked directly to charts, as it is possible to know the 'tune' but to be singing the wrong 'words'.

- Times table charts should be displayed in a focused way: e.g. the 5s and 10s might be displayed in the same week, on another occasion the 2s and 4s, and so on.

- Patterns in individual tables should be explored. This exploration can be for different attributes. For example, the 2s always have a product that ends in 2, 4, 6, 8 or 0, and the order is always the same. The 5s always end in 5 or 0. The product of the 9s always sums to 9. If you wish to extend beyond 10 times then the 11 times table has many interesting features.

- Multiplication squares can be given and the missing numbers found. There are multiplication squares on **Copymaster 69**. An example is given below. Patterns can then be looked for in reinforcement of the multiplications used.

×	**3**	**4**	**5**
3	9	12	15
4	12	16	20
5	15	20	25

The products appear
in the square
Thus 5 × 5 = 25

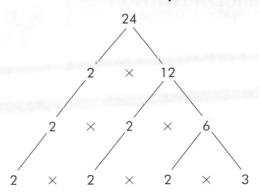

- The construction and exploration of a 10 × 10 multiplication triangle can be very helpful. Not only does it reinforce known facts but will also indicate to the children which of the multiplication facts are the difficult ones to remember. A 10 × 10 multiplication triangle is set out on **Copymaster 28**. Get the children to shade or lightly colour squares that show the products of the 2× then the 5× and 10× tables. Then look for and colour or shade the square numbers. Now colour or shade the 3× and 4× tables. This should leave 6 squares not shaded or coloured. These often prove to be the hardest to remember – but there are only 6 difficult ones.

- Try using known multiplication facts to mentally multiply by 20, 30, 50 and 100. This could be extended into hundreds if so desired.

- Using products that appear in more than one table develops ideas about factors and common factors. Move this on to working out the products of given quantities using prime factors, e.g. 16 has the factors 1, 2, 4, 8, 16; of these only 2 is prime (we usually do not include 1 as a prime); 16 can be generated from its prime factor by 2 × 2 × 2 × 2. Another example is shown opposite.

Assessment

Child's performance	Teacher action
Knows most of the tables to 10 × 10	Give some of the exercises above which provide opportunities to make patterns
Knows all the tables to 10 × 10	Extend the challenges to include the exercises which go beyond 10 × 10
Uses multiplication facts in solving mental number problems	Resource with a range of number challenges which continue to extend the child's use of multiplication facts

Homework

Any of the suggested exercises could be given as homework. The intentions could be revision and rehearsal, or investigation.

Multiplication

Learning targets

On completion of this theme the children should be able to:

1 ➤➤ explain multiplication as repeated addition

2 ➤➤ use known multiplication facts to solve problems

3 ➤➤ solve single-digit by two-digit multiplication problems.

Before you start

Subject knowledge

When dealing with positive whole numbers, multiplication can be seen as a quick way of carrying out the addition of equal-sized groups. Also multiplication is both commutative and associative. It is important children understand these rules. The fact that 3×12 equals 12×3 is vital to the competent manipulation of number, as is the realisation that order does not matter so that $4 \times 3 \times 5$ can be worked out as $(4 \times 3) \times 5$ or $4 \times (3 \times 5)$ or even as $(4 \times 5) \times 3$, and so on. The children will be used to seeing problems set out horizontally across the page. Moving from horizontal to vertical presentation needs careful handling for several reasons. Place value and the positions of digits in the correct columns are clearly important. Also, we are asking the children to read and work from right to left where previously their experience has been left to right.

Note: children need calculators with an iterative function for the following lessons.

Previous knowledge required

Multiplication tables, addition.

Resources needed for Lesson 1

Counters, cubes, or number apparatus such as Dienes® or Cuisenaire®, calculators.

Resources needed for Lesson 2

Counters, cubes and other apparatus as necessary, calculators, Copymaster 29.

Resources needed for Lesson 3

These can include any or all of the resources used in Lessons 1 and 2.

Teaching the lessons

Lesson 1 ①

Key questions

How many is this?
How did you get that answer?
What connection can you see between addition and multiplication?

Vocabulary

Multiply, multiplication, repeated addition.

Introduction | 10 min |

▨ Explain to the whole group that there are 2 parts to the work they are going to do today. They will be using apparatus to work at a range of addition problems, and they will be using calculators. To start, get the children to tackle these sorts of problems:

Make 36 in equal groups of counters (or the apparatus you have). How many ways can they make this? Try now with 15 and if there is time choose another number.

Activities | 35 min |

⠿ Children work with the counters to try and work out ways of grouping, to get the numbers you have asked them to give. It is important that they group them in equal groups. Say, e.g., if they were making 12, then 3 add 3 add 3 add 3, or 4 add 4 add 4, or 6 add 6, and so on would be perfectly acceptable. Bring children back together and share results. Point out the fact that you can add the groups in the way that we have been doing but it might be quicker if we do something called multiplication. Tell them that multiplication is repeated addition. So, e.g., 2×3 means 2 lots of 3 added together (3 add 3), and 3×4 is 4 add 4 add 4. When you are happy with their confidence, give out the calculators.

Demonstrate that if you press, e.g., the number 2 on a calculator followed by the addition sign and then the equals sign and then you keep pressing the equals sign, the calculator display will show the increasing numbers 2, 4, 6, 8 and so on. Let them try this for any numbers they choose. Also, e.g., if the children want to enter $12 + = = =$ and so on, they will be working through the repeated addition of the 12 times table.

Closing the lesson `10min`

Review all that has been done but firstly what the children have just been doing with calculators. This is in order to reinforce the idea of repeated addition with whole positive integers.

Assessment

Child's performance	Teacher action
Can produce sets of equal groups to make the totals required	Move on to next lesson but give more practice on times tables work
Clearly understands multiplication is the same as doing repeated addition	Move on to next lesson
Can explore extensively work done using calculators	Give some calculator challenges which involve the associative law and move on to next lesson

Lesson 2 ②

Key questions

Does it matter which order you do this in?
How did you do this problem?

Vocabulary

Multiplication, total, order, calculation.

Introduction `15min`

Do a quick times table exercise with the class if it is appropriate. Then show them that it is possible to write down problems with more than one multiplication in them. Explain that there is a rule for the order you do the calculation in where there are brackets – the part in the brackets is done first. For example $(3 \times 4) \times 2$ means that we multiply the 3 and the 4 before we multiply all of that by the 2. Give some examples.

Activities `35 min`

Give out **Copymaster 29** and ask the children to try out the problems. They can use calculators to check their answers.

Closing the lesson `10min`

Go through some of the solutions. Make it clear that if someone does the problem in a different order, that is fine provided that the answer produced is correct. Reinforce, e.g., that 3×5 equals 5×3 and that $6 \times 2 \times 8$ can be done in a variety of ways.

Assessment

Child's performance	Teacher action
Works through the first two sections of the copymaster	Work further on commutativity
Completes all the copymaster but finds the last part quite challenging	Work further on associativity
Has no problem with the lesson	Move on to next lesson

Lesson 3 ③

Key questions

What does the 3 mean in 36?
Why do we have to organise our work so that the numbers do not get muddled?
How can we check our answers?

Vocabulary

Digit, total, multiply, multiplication.

Introduction `20min`

Do some times table work. Then set out an example of two-digit by one-digit multiplication (e.g. 21×6). Do this horizontally and invite suggestions as to how we might solve the problem. Work through this example and one or two more vertically, talking through the multiplication of units and 10s and the meaning of 0 in the children's working out. Bear in mind that there are a number of ways of working vertical calculations. Give the children a range of problems to tackle. Include at the end of these some that involve two-digit by two-digit problems such as 23×10, 23×12, and so on.

Activities `20min`

Children work through the problems you have given them. If some get well ahead, see whether they can come up with a method for checking the answers without using calculators.

Closing the lesson `10min`

Get the children to disclose their approaches and discuss possible errors. These include missing out the 0 or putting it in the wrong place, not lining up the columns, and misreading or misremembering the times table.

Assessment

Child's performance	Teacher action
Completes only a few of the problems	More practice needed on single and then on single-/two-digit numbers
Completes all of the problems	Give more practice of single-digit by two-digit multiplication problems
Completes all of the problems and develops a self-checking approach	The learning targets for this theme have been met

Homework

Encourage the children to use calculators to explore the idea of multiplication as repeated addition. Select some examples of two-digit by one-digit multiplications and invite the children to solve them. Get the children to produce as many ideas as they can on where multiplication might be used in everyday life. Many of these will be to do with shopping, e.g. how much are 4 cans at 24p each?

Division

Learning targets

On completion of this theme the children should be able to:
1 ➤ explain the meaning of division
2 ➤ use known facts in division problems
3 ➤ divide two-digit numbers by a single digit.

Before you start

Subject knowledge

Children know a lot about division in the sense of sharing. There are, of course, many ways of interpreting the idea of sharing, e.g. all children will know about sharing pencils or an eraser, but in mathematics we are interested in sharing in the sense that it means: how many will be in a given number of sets? e.g., how many people can share 10 sweets equally? The other main way of thinking about division is grouping. Here we might say, e.g., there are 5 people, how many sweets can each of them have? Of course, this leads us on eventually to consider remainders.

Previous knowledge required

Subtraction, addition, multiplication, especially multiplication tables.

Resources needed for Lesson 1

Counters, cubes, calculators.

Resources needed for Lesson 2

Counters, cubes, calculators, Copymaster 30.

Resources needed for Lesson 3

Base 10 apparatus such as Dienes®, calculators.

Teaching the lessons

Lesson 1 ①

Key questions

How many times did you take the number away to get to 0?
What is, say, 12 divided into 3 groups?
What is, say, the number of groups you could have if you divide equally into 12?
How many groups can you make sharing, say, 12 equally?

Vocabulary

Division, groups, remainder, repeated subtraction.

Introduction 10 min

Write the number 12 on the board. Ask the children what happens if you take 3 away repeatedly. Try again with 15 taking away 5, and 16 taking away 4. Talk about how this might connect with what they know about times tables.

Activities 35 min

 Get the children to work in pairs or small groups to produce ideas starting with numbers that they choose and taking away numbers that they also choose. Go round the class, talking to them, and looking at what they are doing. Note which children can actually come up with remainders. Bring the children back together and show them the memory button on a calculator. If they put a number into the memory and then repeatedly subtract another number, do they get back to 0? How many pushes of the button do they need?

Closing the lesson 10 min

Use the children's ideas to explore what division might be. Take their ideas and revise these with them, reinforcing those that are key in understanding division.

Assessment

Child's performance	Teacher action
Can carry through work with physical objects but clearly not comfortable with the ideas	More work needed using practical apparatus
Can do work with counters with confidence	Move on to next lesson but monitor progress
Can produce answers without using counters or the calculator	Invite a child to create some problems for a class database of division challenges; move on to next lesson

Lesson 2 ②

Key questions

How did you solve this?

Can you explain to me what division is?

Can you see any connection between multiplication and division?

Vocabulary

Division, groups, shares, remainders.

Introduction | 15 min |

 Start with a times table exercise and then choose some of the multiplications and put them on the board. For example, use $6 \times 2 = 12$. Then ask what $12 \div 2$ might be. And what about $12 \div 6$? Try others until you feel that the children are making the right connections. Remind them of the fact that division can be seen as repeated subtraction. Then give them **Copymaster 30**.

Activities | 25 min |

 Children work through Copymaster 30. Move around the class asking questions, particularly about how they are tackling the given problems.

Closing the lesson | 15 min |

Put up some more problems of a similar type to those on the copymaster, on the board, and get the children to explain how they can be solved.

Assessment

Child's performance	Teacher action
Participates but shows considerable hesitation	Continue practice with times tables and give more calculator practice
Completes work satisfactorily	Move on to next lesson but give more practice in linking multiplication, division, addition and subtraction
Clearly has a good grasp of the ideas	Move on to next lesson

Lesson 3 ③

Key questions

What are the important things to remember in division?

Vocabulary

Division, vertical, horizontal.

Introduction | 20 min |

Remind the children of the work they have done on times tables and division tables. Set out some examples to help them recall this. Explain that we can use this knowledge to help us deal with larger quantities. Put up an example of a two-digit divided by a one-digit problem, e.g. $56 \div 4$. Explain that it is quite useful to arrange this problem in a particular way when doing division. Work through an example. Then try another one such as $42 \div 3$. When you feel the children are happy with the process, give them a set of similar examples to work through.

Activities | 30 min |

 The children work through the examples you have given them while you circulate, ask questions and listen to their thoughts on the process of doing division. Encourage them to use calculators to check their answers.

Closing the lesson | 10 min |

Write up on the board another example, which this time will have a remainder. Work through this. In so doing, you will reinforce what has been practised in the lesson but will also alert them to the fact that this learning can be extended.

Assessment

Child's performance	Teacher action
Follows the explanation but makes errors in practice	Give more practice in class where you can help
Completes the task satisfactorily	The learning targets for this theme have been met
Is clearly confident with all including the remainders	Move on to a theme involving remainders

Homework

Get the children to produce times tables and division tables for the same number. For example, $4 \times 1 = 4$, so $4 \div 4 = 1$. Set some problems involving numbers of things, money, and measurement where the children have to divide in order to get the answer. Start with those that have whole number answers. For those that have a good grasp of the ideas, extend these to whole numbers and some left over. Get the children to choose an even number, such as 12, and an odd number, such as 15, and invite them to show how many ways these numbers can be equally divided. Let the children experiment with calculators and the division key to see what they discover.

Remainders

Learning targets

On completion of this theme the children should be able to:

1 ➤➤ handle simple division in which there are incomplete sets or groups
2 ➤➤ use pencil and paper methods to calculate simple divisions with remainders
3 ➤➤ round to the nearest 10, 100 and 1000.

Before you start

Subject knowledge

There are a number of factors to consider when doing division problems in which there are not whole number answers. For example, there are aspects of the way in which we expect children to record their answers. They need to be able to operate an appropriate written algorithm, taking proper account of place value. They also need to know the conventions for writing out their answers. In the early stages, an 'r' is commonly used to indicate the quantity which is the remainder. It may be the case that the remainder is a larger number than the whole number answer so it is necessary for the children to appreciate that in, for example, 19 ÷ 5, the answer of 3, remainder 4 means 3 lots of whole 5s and 4 left over. Also using the written algorithm we often want the children to use vertical presentation. Whilst this may help with place value, it can be awkward for some children for we are asking them to read the numbers vertically too. Finally, the children need to come to appreciate that the size of a remainder is useful in thinking about the nearest whole number. Rounding up and down is part of their understanding of division.

Previous knowledge required

Simple division, multiplication tables.

Resources needed for Lesson 1

Counters, spinners (such as on Copymaster 61) or blank dice – enough of these for 4 each between small groups.

Resources needed for Lesson 2

Counters, as necessary.

Resources needed for Lesson 3

Number cards made up from Copymaster 31 (make up from several copies of the copymaster if required), calculators, if needed, for extension work.

Teaching the lessons

Lesson 1 ①

Key questions

What do these total?
What are the easy ones to do?
How many halves?
What remains?
What is left over?

Vocabulary

Division, remainder, total.

Introduction 10min

On the board, write a number between 2 and 7 inclusive. Then write 3 numbers beside the first one that come between 7 and 12 inclusive. With the children, total the 3 larger numbers. Ask them whether the first, smaller number will divide equally into this total. If not, then how many remain? Explain that they will have spinners or dice, on 3 of which the numbers 7, 8, 9, 10, 11 and 12 appear. On the fourth spinner or dice, there are the numbers 2, 3, 4, 5, 6 and 7.

Activities 20min

Working in small groups or pairs, children have to make up some division problems using the dice or spinners in the same way as shown on the board, and solve them. Tell them that they can use counters to help, if they wish. Check round for those who might be at a loss what to do about remainders.

Closing the lesson 15min

Listen to some of the children's problems and solutions. Use these to go over the ways you would like the written algorithm to appear. Get the children to suggest which of the problems was easiest and why. They may suggest odd and even number or multiples of 5 or 10, and so on.

Assessment

Child's performance	Teacher action
Has to use the counters throughout	Give more work on the division of whole number problems
Participates fully and gives appropriate answers	Move on to next lesson
Can do many of the problems mentally	Move on to next lesson but consider giving the opportunity to extend understanding by developing some own division problems using calculators

Lesson 2 ②

Key questions

What do we have to remember when doing written division problems?
How do we know what part of the answer is the remainder?

Vocabulary

Division, remainder.

Introduction
⬛ 15min

Revise what the children should now know about multiplication and division. This will include times tables and their relationship to division tables, as well as the written algorithms which you have been teaching them. Ask them to explain what clues they look for when tackling division problems. For example, does it help to remember the final digit in tables such as 5×, 10×, 2× and 4×, when making a judgement whether the division will produce a whole number solution? Give the children some exercises you have chosen which will let them practise written division problems.

Activities
⬛ 25min

Let the children work through some exercises that you have chosen. These should include whole number problems as well as those with remainders. Have available some more challenges for those who race through your first selection.

Closing the lesson
⬛ 15min

Bring the children together and select a sample from the problems that have different features and offer different sorts of challenges. Go over these to reinforce what the children should now have learnt.

Assessment

Child's performance	Teacher action
Has only partial success	More work needed with simple division with whole number problems; consider getting out base 10 apparatus and again working through the ideas practically
Completes the problems	Move on to next lesson
Completes the problems and tackles the extension activities	Move on to next lesson

Lesson 3 ③

Key questions

Why is 10 so important to our number system?
What is this to the nearest 10, 100, 1000?

Vocabulary

Number vocabulary, rounding, round up.

Introduction
⬛ 20min

Start by asking what the children know about 10 and get suggestions as to why it is so important. Use the opportunity to reinforce place value, times 10, times 100 and times 1000, as well as dividing by 10, dividing by 100 and dividing by 1000. Tell them that now they are going to round up numbers to the nearest 10, 100 or 1000. Show them the rule that says that a number ending with 5, 6, 7, 8, 9 is rounded up to the nearest 10 and a number with a 1, 2, 3, 4 is rounded down. Put examples on the board from the following ranges: 30–40 and ask them to round to the nearest 10; 400–500 and ask them to round to the nearest 100; 6000–7000 and ask them to round to nearest 1000.

Activities
⬛⬛ 25min

You will need to prepare number cards from **Copymaster 31**, Working in pairs, children must sort their cards and agree where they go after rounding (e.g. in a '100 pile' or '200 pile'). Those children who get on well could be invited to take, e.g., the 151 card and write it to the nearest 10 as well as to the nearest 100.

Closing the lesson
 15min

Choose some of the examples you have seen around the classroom as the children have worked and use these to go over the rules again.

Assessment

Child's performance	Teacher action
Copes with some of the challenges	Give more practice time with support
Does all with a good degree of success	The learning targets for this theme have been met
Completes all and goes on to find nearest 10s and 100s	The learning targets for this theme have been met

Homework

Give calculator challenges where the children produce numbers that interest them and try to work them out to the nearest 10, 100, 1000 or to the nearest 10,000, 100,000 and so on. Give some division examples based on measurement, including money. Invite the children to identify board games they know, where you have to move a counter. What happens if your last throw is too much? Are there rules for this? Are there home rules for this? Could they invent a game that uses division and remainders?

Long multiplication

Learning targets

On completion of this theme the children should be able to:

1 ➡ multiply mentally by 10, 100, 1000, 20 and 50
2 ➡ undertake long multiplication reliably using at least one method
3 ➡ employ a range of approaches to long multiplication.

Before you start

Subject knowledge

There are three points to consider here. The core ideas to embrace when working on long multiplication are to do with place value, the need for flexibility in approach and when to use the calculator.

In respect of place value, it is of real importance that children do not use the phrase 'add a nought'. Adding a nought has no effect, as we know. The importance in long multiplication is that it holds a place, whether it is in the units, 10s, 100s or other columns.

It is best if a variety of written algorithms are explored as all the evidence is that, at this age, flexibility in mental arithmetic can be inhibited by too restrictive an approach. To undertake long multiplication, children do need a sound grasp of their multiplication table facts. It does not matter whether multiplying is done using the 100s first,

then the 10s, then the units, or the other way around, as long as children are confident with place value.

Children should be allowed to use calculators appropriately. Certainly they should be used for large quantities but, in this case, estimation should come first. Work on multiplying by 10, 100, 1000 should be of value here, as is the work done on rounding up and rounding down.

Previous knowledge required

Multiplication tables, simple written multiplication.

Resources needed for Lesson 1

No special resources required.

Resources needed for Lesson 2

Copymaster 32.

Resources needed for Lesson 3

Copymaster 33.

Teaching the lessons

Lesson 1 ①

Key questions

What is … times …?

Vocabulary

Number vocabulary, multiplication, product.

Introduction 20min

The work in this lesson is concerned with mental arithmetic using multiples of 10. The children should be used to mental arithmetic challenges as a regular part of their mathematics work. Here we are trying to explore and reinforce the outcomes of 10×, 100× and 1000× and further. Mental multiplication using 20× and 50× should also be attempted. Get the

children together, having prepared a set of challenges, working up from 10× to the other quantities. Put some of the challenges on the board and talk through what is happening.

Activities 20min

Working in pairs or small groups, the children have to prepare their own rapid-fire challenges together with answers. They should produce at least 5 questions and solutions.

Closing the lesson 20min

Get the children back together. Arrange the groups so they are sitting together and then each group can fire a question for the rest of the class to have an attempt at.

Assessment

Child's performance	Teacher action
Child rather reticent or hesitant	Praise where possible; take time later to explore any gaps in times table knowledge
Participates fully	Praise and move on to next lesson
Participates with clear enthusiasm and understanding	Praise and invite ideas for the next time you do this lesson; move on to next lesson

Lesson 2 ②

Key questions

What does the 'zero' stand for?
Why do we need to be precise about where we write the numbers?

Vocabulary

Digit, long multiplication.

Introduction

Start with a two-digit by one-digit example on the board. The layout below suggests a variety of ways of portraying a multiplication. When you feel the children are secure in what you are explaining, then work through a two-digit by two-digit example. Do more examples until you sense the children's confidence.

```
  32 ×          32 ×
   9            1 9
 ───          ───
 288          288
  1

  32 ×          32 ×
   9             9
 ───          ───
  18          270
 270           18
 ───          ───
 288          288
```

Activities

Give out **Copymaster 32** on which there are 3 examples of long multiplication of two-digit by two-digit numbers, plus some to try. At the end of the copymaster are a few three-digit by two-digit problems for the adventurous ones. The children should work on the copymaster with your help and support

Closing the lesson

Hold a class discussion. Ask questions: What was easy? What was hard? What was confusing, if anything? Use the children's responses to fuel the understanding of others.

Assessment

Child's performance	Teacher action
Makes some progress	Encouragement is needed plus further practice
Completes the two-digit by two-digit problems satisfactorily	Move on to next lesson
Engages with all the problems	Move on to next lesson

Lesson 3 ③

Key questions

Why and how do you think this method works?
How do these different ways compare?

Vocabulary

Long multiplication, method.

Introduction

Give out **Copymaster 33** and discuss the examples given there, explaining what the processes are. Work through 1 or 2 problems using each method, on the board.

Activities

Invite the children to tackle the problems on the copymaster using the different methods as they choose. Encourage them to discuss the merits of each method and, if there is time, ask them to try the same problem using any other method they can think of.

Closing the lesson

Get individuals to show some of their attempts to the class.

Assessment

Child's performance	Teacher action
Develops one of the methods	Check on progress and subsequent related work
Develops a range of approaches	The learning targets for this theme have been met
Develops a range of approaches and makes further suggestions about them	The learning targets for this theme have been met

Homework

Give multiplication challenges either produced by you or in other resource material. Ask the children to list all the coins in circulation and say what they need to be multiplied by to make £1, £5, £10, £20 and £50.

Long division

Learning targets

On completion of this theme the children should be able to:

1 ➡➡ mentally divide by 2, 5, 10 and use base 10 materials to explain division
2 ➡➡ work out given problems using short division
3 ➡➡ work out given problems using long division.

Before you start

Subject knowledge

The available evidence suggests that children who have had a good grounding with practical apparatus such as base 10 materiasl find the development of written algorithms for short and long division less of a problem than those who have not had this experience. In recent years, there has been a blurring of the old distinction between short and long division. However, we believe that teaching both is important in support of the practical, shorthand methods that adults develop in their everyday lives. The key ideas which underpin written divisions of both sorts are place value and repeated subtraction. Below, an example of division by repeated subtraction is given.

$15 \div 4$

$15 - 4 = 11$

$11 - 4 = 7$

$7 - 4 = 3$

3 '4s'   remainder 3

Whilst this is perfectly acceptable as a written algorithm, there is an expectation that what are seen as more traditional methods can be

employed. Whilst, for large quantities, use of a calculator makes more sense, providing we estimate first, it is still useful to have the skill of dividing two-digit, three-digit and more than three-digit numbers. In addition to the listed vocabulary, there are some words that you may choose to use, depending on your judgement. These are *dividend* – the amount to be divided, *divisor* – the number that is being divided into the dividend, *quotient* – the answer to a division calculation.

Previous knowledge required

Division as repeated subtraction, division tables, multiplication tables.

Resources needed for Lesson 1

Base 10 materials, e.g. Dienes® apparatus.

Resources needed for Lesson 2

Examples for solution by short division including Copymaster 34. Calculators should be available if you wish to check answers.

Resources needed for Lesson 3

Examples for solution by long division. Again, calculators can be available for checking if you wish.

Teaching the lessons

Lesson 1 ①

Key questions

What does zero represent?
What patterns can you see that can help us in dividing by certain numbers?

Vocabulary

Divide, division, place value, zero.

Introduction [20min]

▓ With the class, continue the practice they will have had of mental mathematics. In this case, concentrate on division and pose problems where dividing by 2, 5

and 10 are important. Using base 10 apparatus and columns on the board, explain the process of division with the help of the apparatus. Use examples such as $572 \div 22$ and $378 \div 18$. Now use examples such as $403 \div 13$ and $720 \div 36$ to review the role and importance of the use of 0.

Activities [25min]

 Set out a range of division problems for the class to attempt. They should use base 10 apparatus to help them do this.

Closing the lesson [15min]

▓ Take some examples used and work through them with the help of the class.

Assessment

Child's performance	Teacher action
Fine with mental arithmetic but has some problems with the division	Give more practice with base 10 materials
Works satisfactorily through all the problems	Move on to next lesson
Works very competently through all the problems	Move on to next lesson

Lesson 2 ②

Key questions

Why do we put the numbers where we do?
Why do we use these symbols?

Vocabulary

Divide, short division, carrying, borrowing.

Introduction

In this lesson we are concerned with work on short division. Here are some examples:

$$195 \div 15 \qquad 272 \div 18$$

Set out these and similar examples on the board progressively and work through them with the class. Ask questions about the process. Why do we use the particular signs? Where is the answer? How do we remember the parts of the solution as we work through?

Activities

Give the children **Copymaster 34** to work through on their own or in small groups. Work with the groups, checking on their times tables, the layout they are using and picking up on any particular queries.

Closing the lesson 15 min

Ask the children what, if anything, caused them problems. Work through a few of the challenges they have had, collecting their work.

Assessment

Child's performance	Teacher action
Gets part way through	Find time for more practice, following further explanation by you
Completes the problems with some errors	Move on to next lesson
Completes the problems with no errors	Move on to next lesson

Lesson 3 ③

Key questions

Where do we place the different numbers and why?
What symbols do we need to use?
Why is it important to keep our work in columns?

Vocabulary

Divide, long division, carry, borrow.

Introduction

As with the lesson on short division, the intention is to provide formal practice in at least one of the written algorithms for long division. However, it is worth developing your algorithm step by step. Examples of these steps are given below for one algorithm.

596 ÷ 14

```
      42 r8
14)596
   −140    (14 × 10)
    456
   −140    (14 × 10)
    316
   −140    (14 × 10)
    176
   −140    (14 × 10)
     36
   − 28    (14 × 2)
      8
```

Can lead to:

```
      42 r8
14)596
   −560    (14 × 40)
     36
   − 28    (14 × 2)
      8
```

Use this example, or another you choose, and work through the steps on the board. Discuss with and question the children at each stage, concentrating on place value, 0 and the need for careful writing and columns. Provide examples for the children to work through on their own or in small groups.

Activities

Children work through your examples. As with short division, check on such things as times tables, layout and the use of 0.

Closing the lesson 15 min

Ask the children where they got stuck or had to think particularly hard. Work through a few examples and then collect in the children's work.

Assessment

Child's performance	Teacher action
Gets part way through	More practice and go back to earlier work if necessary
Completes the problems with some errors	More practice required
Completes the problems with no errors	The learning targets for this theme have been met

Homework

Make up or extract from available resources problems in written form. For example: A garage wants to fill all of one kind of car with fuel. Each car holds 22 litres and the garage has 335 litres of fuel. How many cars can they fill up?

Multiplying decimals

Learning targets

On completion of this theme the children should be able to:

1 ➡➡ multiply whole numbers by a decimal number
2 ➡➡ multiply simple decimals
3 ➡➡ explore decimals further using a calculator.

Before you start

Subject knowledge

There are many areas of everyday activity where we meet decimals. These include measurement of length, weight and money, though our currency can create some misunderstandings in decimal computation. These misunderstandings arise from the way we 'say' amounts of money. One cannot have more than 2 decimal places. Commonly we talk about, for example, £1.89 as 'one pound eighty nine' whereas, in decimal work, we would denote 1.89 as 'one point eight nine'.

The decimal point itself can be a source of great anxiety for children. They worry about where it should go in an answer. There are two responses to this. We can get rid of the decimal point by multiplying by a multiple of 10. For example, in 23×1.2 we could multiply both by 10 and obtain 230×12. We then need to divide the result by 100 (10×10) to get our answer. However, this can, in our experience, cause confusion in the early stages though it will be important later on. So, we would advocate the use of estimation. In the example given, 23×1.2, this is going to produce something

a little more than 23×1 so examination of the product will tell us where the decimal point should be placed. In order to become confident in handling decimals, the children need to understand place value and that, just as we have 100s, 10s, units, we can have tenths, hundredths and so on. The decimal point indicates the point of change between the units and the tenths.

Previous knowledge required

Multiplication tables, long multiplication, calculator use, the addition and subtraction of decimals.

Resources needed for Lesson 1

Copymaster 35, calculators

Resources needed for Lesson 2

Calculators, some problems involving the multiplication of decimals by decimals with one digit after the decimal point.

Resources needed for Lesson 3

Calculators.

Teaching the lessons

Lesson 1 ①

Key questions

Why do we need a decimal point?
What is happening in terms of place value?
What happens when we multiply a decimal by 10, 100, and 1000?

Vocabulary

Decimal point, place, estimation.

Introduction | 15 min |

▓ Give the class calculators, at least 1 between 2 children. Get them to enter 0.5, then to multiply this by 10. Put the answer on the board. Now enter 0.25 and again multiply by 10. What happens? Discuss this with

the class, putting up 10s, units and tenths columns on the board. Get the children to try more problems for themselves.

Activities | 25 min |

👥 The children should make up some of their own examples, using the calculator, and recording their results clearly. If any suggest that they would like to multiply by 100 or 1000, then encourage this.

▓ After a few minutes, get the class back together and show them how you might use long multiplication to solve problems involving whole numbers and decimals. Start with examples already used then show an example that does not involve 10 or a whole multiple of 10. Give the children **Copymaster 35** to work out for themselves using long multiplication and then checking the answers with the calculator.

Closing the lesson [10 min]

 Bring the class back together and choose individuals to explain the examples that have been attempted. Reinforce the approach and briefly demonstrate that estimating an answer before we start is good practice and this will reduce the possibility of putting the decimal point in the wrong place.

Assessment

Child's performance	Teacher action
Can accomplish the first part of the task but not clear on later examples	Consider giving more practice using the calculator and revise long multiplication
Completes all parts of the task	Move on to next lesson
Completes all parts of the task and extends with own ideas	Move on to next lesson

Lesson 2 ②

Key questions

Where does the decimal point go?
How can we get an estimate of the answer?
Why is estimation important?

Vocabulary

Decimal point, place, multiplication, estimation.

Introduction [20 min]

 Give out calculators, at least 1 between 2 children. Write the table shown below on the board.

×	24	2.4	0.24	0.024
5				
0.5				
0.05				
0.005				

Get the children to estimate answers for the empty cells and then, using the calculator, the correct answer. Look at and discuss the patterns that are in the table. Discuss what is happening in respect of where the decimal point is located. Then put up an example such as 3.2 × 1.2 on the board and work through this. Give out the problems you have previously prepared.

Activities [20 min]

 Children should work through the examples you have given. In each case, get them to write an estimate before carrying out the calculation. They can check with the calculator afterwards.

Closing the lesson [10 min]

 Put up 1 or 2 of the answers and work through them starting with estimation. Then work through the written multiplication and locate the decimal point.

Assessment

Child's performance	Teacher action
Has problems with locating the decimal point	More work on estimation and, possibly, long multiplication
Works through the problems satisfactorily	Move on to next lesson
Finds the work straightforward and can extend it with own ideas	Consider child creating a database of similar problems for class use at a later date; move on to next lesson

Lesson 3 ③

Key questions

Why does the decimal point go here?
What happens if …?

Vocabulary

Decimal point, multiplication, estimation.

Introduction [15 min]

 Remind the children of all the work they have done on multiplying decimals. Emphasise estimation, long multiplication, place value and the meaning of the decimal point. Give out calculators and tell the children that the task is to make up some multiplication problems with which to challenge each other at the end of the lesson. They do not have to stick to 1 decimal place.

Activities [25 min]

 The children work in small groups to create at least 5 good examples. Encourage them to think about multiplying by less than 1, more than 100, and so on.

Closing the lesson [15 min]

Elicit at least 1 example from each group. The rest of the class has to solve these, using calculators as necessary. Use the examples to discuss again the principles behind the multiplication of decimals.

Assessment

Child's performance	Teacher action
Makes a contribution but is still clearly lacking in confidence	At a later stage, give more opportunities for multiplication practice
Makes a contribution and has good insights	The learning targets for this theme have been met
Gives a lead, producing many good examples	Engage in the compilation of the class database of decimal multiplication problems

Homework

Give some multiplication of money challenges as a means of practising multiplication of whole numbers by decimals. Examples are a take-away meal for 65 at £2.95 each, and 3 tickets for the cinema at £6.85.

Dividing decimals

Learning targets

On completion of this theme the children should be able to:

1 ➡➡ divide whole numbers where the result is a decimal
2 ➡➡ make connections between division of whole numbers and division of decimals
3 ➡➡ solve basic long division problems involving decimals.

Before you start

Subject knowledge

These days division involving decimals is commonly carried out using calculators, especially when working with large quantities. However, children do need to know how the division of decimals works. There are a variety of algorithms available but 2 features must be considered as being at the heart of any method used; the importance of estimation in determining the placing of decimal points, and the understanding of place value. It is best to avoid statements like 'add a nought', as this is mathematically unsound. Also, we need to be careful of undertaking division where we talk about, for example, 'going into 2' when we actually mean 20 or 200.

It is also important that children come to appreciate that the larger the divisor, the smaller will be the answer; and, conversely, the smaller the divisor, the larger the answer. It can be useful to talk of how many of the divisor is needed to make the dividend. So, e.g., in 12.5 ÷ 2.5, it takes five 2.5s to make 12.5. Finally, in real life problems, it is usually the case that decimals do not work out conveniently to 1 or 2 places. Recurring decimals are commonplace. Children should be made aware of this and you can use this to develop further their ideas about significant figures and decimal places.

Previous knowledge required

Division, long division, multiplication of decimals, use of calculators.

Resources needed for Lesson 1

Copymaster 36, some decimal long division problems, and calculators.

Resources needed for Lesson 2

Calculators.

Resources needed for Lesson 3

Copymaster 37.

Teaching the lessons

Lesson 1 ①

Key questions

Where should the decimal point go?
How do you know where to put the decimal point?

Vocabulary

Long division, decimals, decimal point.

Introduction 20min

▦ Go over a long division problem on the board to remind the class of one of the ways in which such a problem can be tackled. Choose an example such as 276 ÷ 12 that gives a whole number answer. Now do one with a remainder. Try 78 ÷ 12. Explain that, instead of a remainder, we can use a decimal answer. Work out 78 ÷ 12 alongside the example on the board to show how 6.5 can be produced.

Activities 25min

🔲 Give out **Copymaster 36** which gives some practice examples and then some challenges. In undertaking the challenge, the children will find it quite difficult to find examples that have only 1 decimal place in the answer. This fact can promote a useful discussion about decimals generally, both in the activity period and at the end of the lesson. The children need to try to replicate the processes you have explained to them. Give out calculators for children to check their progress and take on the challenge in the second part of the copymaster.

Closing the lesson 15min

▦ Choose some examples and complete them on the board. Use the opportunity to reinforce ideas on estimation, place value and decimals in their connection with remainders.

Assessment

Child's performance	Teacher action
Can tackle the first part of the copymaster but finds the rest harder	Give some more practice and, if necessary, revise division, remainders and decimal fractions
Can complete all of the task	Move on to next lesson
Completes all of the task and raises new and appropriate questions	Move on to next lesson

Lesson 2 ②

Key questions

Why does this happen?

Vocabulary

Decimals, decimal point, place value.

Introduction 15min

 Give out the calculators, at least 1 between 2 children and ask the children to key in 30 ÷ 6. They should know the answer to this without the calculator, but do it with the calculator because of the next step. Now get them to key in 3 ÷ 0.6. See what they think. Try some more of this type and then set out an example on the board, pointing out that if we multiply both the dividend and the divisor by 10 then we simplify the problem.

Activities 20min

 Ask the children to produce at least 5 examples of the type you have shown, including writing them out as division problems. A typical response should look like this: 6 ÷ 0.2 → 60 ÷ 2 → 30 = 2 into 60, and 4 ÷ 0.8 → 40 ÷ 8 → 5 = 8 into 40.

Closing the lesson 15min

 Extend what the children have been doing by putting up this sort of problem, 6 ÷ 1.2 and work it through as 60 ÷ 12. Get the children to check with their calculators.

Assessment

Child's performance	Teacher action
Completes task but shows some uncertainty when questioned	Give more practice and consider further work on division of whole numbers
Completes task satisfactorily	Move on to next lesson
Completes task and extends with own ideas	Move on to next lesson

Lesson 3 ③

Key questions

How might we do this?
Why can we put extra zeros?

Vocabulary

Divisions, long division, decimal point, place.

Introduction 20min

 Here we consider a traditional long division algorithm to extend the children's appreciation of the process. It also serves to make them familiar with some of the algorithms that are in common use in the wider society. Put an example on the board, e.g. 14.4 ÷ 4, and work through it, as shown below.

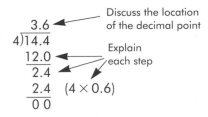

$$\begin{array}{r} 3.6 \\ 4\overline{)14.4} \\ 12.0 \\ \hline 2.4 \\ 2.4 \quad (4 \times 0.6) \\ \hline 0\,0 \end{array}$$

Discuss the location of the decimal point
Explain each step

Give out **Copymaster 37**. Talk through the process, explaining about the tradition for moving figures and the importance of locating the decimal point correctly. Use estimation to check this though the convention is to line up the point with that in the dividend.

Activities 20min

 The children should work through the copymaster. Keep reminding them of what they already know about long division and the importance of estimation to check on the location of the decimal point.

Closing the lesson 15min

Use this opportunity to review what the children have learnt about the division of decimals. The key things are estimation, use of multiples of 10 and the fact that the smaller the divisor the larger will be the quotient. You can demonstrate this using, e.g. 60 ÷ 0.3 and 60 ÷ 0.2. If necessary, use a term like 'makes' here, stating that it takes 20 '0.3s' to make 6 and 30 '0.2s' to make 6.

Assessment

Child's performance	Teacher action
Finds the task difficult	Move away from this work and give more practice on whole number division; consider concentrating on a multiple of 10 approach exclusively
Manages the task satisfactorily	The learning targets for this theme have been met
Has no real problems with this way of doing division	The learning targets for this theme have been met

Homework

Get the children to find a recipe that states how many people it is intended for. Ask them to work out quantities for twice as many, or three times as many, and a third or a half or a quarter as many.

Ratio

Learning targets

On completion of this theme the children should be able to:

1 ➟ make some correct statements about ratio

2 ➟ use fractions as ratios and ratios as fractions

3 ➟ explain some of the attributes of the golden ratio.

Before you start

Subject knowledge

Children have had early experiences of ratio, for example when they have used resources for counting such as Dienes® and Cuisenaire®. They will also know about comparative sizes of similar objects or shapes. Ratio is about comparing 2 quantities. We commonly use *proportion* when talking about ratios in more than one dimension. Ratio is independent of the units of measurement that might be involved. For example, if Ms A gets £600 a week and Mr B £200, the ratio is 600 to 200 or 6 to 2 or 3 to 1. The fact that the currency is pounds sterling is of no relevance, it could just as easily be dollars.

Ratio is implicit in many everyday activities such as cooking, the gearing of machinery, the proportions in a painting and probability. The Ancient Greeks were much preoccupied by the *Golden Ratio* that they used extensively in their art and architecture. It also occurs in many natural objects, e.g. the spiral of sheep and antelope horns and the spirals on pine cones and in sunflower seed heads. The *Golden Ratio* is about 1.618 to 1 and can be found in the Fibonacci series. The Fibonacci series is 1, 1, 2, 3, 5, 8, 13, 21, 34, 55 and so on, each preceding pair of numbers being summed to make the next number, hence 13 + 21 = 34 and 21 + 34 = 55. Dividing say, 55 by 34 gives a ratio 1.618 to 1, to 3 decimal places.

Previous knowledge required

Fractions, equivalent fractions, multiplication.

Resources needed for Lesson 1

Cubes or pegs and peg boards or counters, Cuisenaire® rods or similar materials such as Multi-link®, lamp, balances, Russian dolls and similar objects of different sizes such as toy cars, saucepans, dolls, paintbrushes.

Resources needed for Lesson 2

Squared paper (Copymaster 60), coloured pencils, erasers.

Resources needed for Lesson 3

Calculators, Copymaster 38.

Teaching the lessons

Lesson 1 ①

Key questions

What is the connection between these?
How can we compare these?
What ratio is involved here?

Vocabulary

Comparison, ratio, fraction, proportion, scale, familiar 'size' vocabulary.

Introduction [20min]

Using the objects you have assembled, get the children to make comparisons. Use their statements to start to refine the fact that we can compare using 1, 2 or 3 dimensions for physical objects. Explain that it is possible to compare objects on the basis of a ratio of one to the other. Show them either some Cuisenaire®

or Multi-link® or similar apparatus to illustrate that, e.g. 3 compared with a 6 is 3:6 or 1:2 as the larger is twice the size of the smaller and the converse.

Activities [20min]

 Put on the board a series of ratio statements such as 1:5, 2:3, 5:10, 1:6 and 2:8. Using the apparatus provided, the children have to set out counters, pegs or cubes in these ratios. Some of the children will see that some of your ratios can be simplified, e.g., 2:8 is the same as 1:4. Encourage them to share their insights with others in the class.

Closing the lesson [15min]

Review the work that has been done, selecting some of the ratios and talking through the possible answers. Now use the lamp to produce some shadow effects and try to get the ratio of your hand twice as large, 3 times as large and so on. Use this as a fun activity to reinforce the idea of a ratio of 1:2, 1:3 and so on.

Assessment

Child's performance	Teacher action
Completes task in a literal fashion	Revise work on equivalent fractions
Participates fully and sees that some ratios can be simplified	Move on to next lesson
Participates and suggests other ratios or ideas about ratios	Move on to next lesson; give further challenges such, e.g. the ratio of length to breadth of the classroom or the hall

Able to produce sets of squares which match the ratios and to simplify them	Move on to next lesson; provide further challenges

Lesson 2 ②

Key questions

What is the ratio?
Can you simplify this ratio?

Vocabulary

Fraction, equivalent fraction, ratio.

Introduction $\boxed{15\,min}$

 Draw 2 sets of 6 squares on the board. Rub out 1 of the squares in 1 of the sets and ask what ratio the class thinks the 2 drawings are now in. Work towards 5:6. Do the same with 2 sets of 4 squares, rubbing out 1 in 1 of the sets and work towards 3:4. Now draw 2 more sets of 4 squares and this time; rub out 2 squares in 1 of the sets. Work towards 2:4. Remind them of equivalent fractions in order to get a ratio of 1:2. The smaller set is half that of the larger.

Activities $\boxed{20\,min}$

 Give out squared paper and write up a series of fractions and ratios on the board. These should include, e.g. 2:3, one sixth, one quarter, 2:5, 4:3 and three sixths. The children should produce sets of squares coloured or shaded to correspond to these fractions and ratios. Encourage them to think about how they might simplify particular answers.

Closing the lesson $\boxed{15\,min}$

Using the board, or an overhead projector if available, put up squares in order to illustrate some of the problems the children have been tackling.

Assessment

Child's performance	Teacher action
Reproduces similar patterns but cannot fully explain	Give some work on comparative measures of length and revise equivalent fractions
Able to produce sets of squares which match the ratios	Move on to next lesson

Lesson 3 ③

Key questions

How is this series of numbers developed?
What is the ratio of these numbers?

Vocabulary

Ratio, Golden Ratio, Golden Section.

Introduction $\boxed{15\,min}$

 Put the series of numbers 1, 1, 2, 3, 5, 8 on the board and ask the children how this series is being developed. If necessary, add 13 then 21. When they see the process for the series generation, talk to them about the fact that each pair of numbers can be seen as a ratio. So 1:1 then 1:2 then 2:3 and so on.

Tell them that the series is named after an Italian known as Fibonacci who came from Pisa and whose first name was Leonardo.

Activities $\boxed{30\,min}$

 In small groups, get the children to develop the series and, using calculators, to work out the ratio of pairs of numbers in the series. Get the class back together briefly and ask for their results. They should have got something close to 1.62. Give out **Copymaster 38** and ask the children to find rectangles, measure the length and the breadth, and work out their ratios.

Closing the lesson $\boxed{15\,min}$

Make an overhead projector transparency of Copymaster 38 if possible and use this to identify the rectangles. Get the children to give you their ratios. Where a rectangular part of the structure is in perspective, there may be some distortion. Make the point about the Golden Ratio and its importance to the Greeks.

Assessment

Child's performance	Teacher action
Can complete the series but struggles with ratios both in numbers and on the copymaster	Find time for some individual work using the copymaster
Completes the series and ratios	The learning targets for this theme have been met
Completes the series and ratios and raises further questions	The learning targets for this theme have been met

Homework

Measure some dimensions of rooms at home and calculate the ratio of length to breadth. Get the children to collect some sports league tables and compare the home and away results. What ratios can they find?

Scale

Learning targets

On completion of this theme the children should be able to:

1 ➤➤ talk about the meaning of scale
2 ➤➤ produce a scale drawing
3 ➤➤ interpret scale in relation to maps.

Before you start

Subject knowledge

Scale is encountered in everyday life with children, especially in toys and scale models of cars, trains and aeroplanes. The children will also be familiar with the fact that photographs can give you either scaled-down pictures or enlargements. The use of the camera can also lead on to thoughts about other optical instruments where the experience of scale is important, including the telescope and binoculars. Reduction and enlargement are key ideas that relate to scale and scaling. In accurate scale drawings of plans or maps, there is the notion of a scale factor too. For example, many scale models are constructed on a 1:72 scale. This scaling links, of course, with ratio, proportion and percentages.

Previous knowledge required

Measuring, ratio, use of drawing instruments, the construction and use of line graphs.

Resources needed for Lesson 1

A variety of scale models such as cars, model railway components and a small dolls' house, graph paper, postcards or photographs of buildings, squared paper (Copymaster 60), triangular dotted paper (Copymaster 70), geoboards, elastic bands.

Resources needed for Lesson 2

Measuring tapes and rulers, squared paper (Copymaster 71), plain paper, scissors and card, maps of the school buildings, drawing instruments.

Resources needed for Lesson 3

Copymaster 39, maps of an area where walking is possible, part of an Ordinance Survey (OS) map or text books in which there are extracts of such maps.

Teaching the lessons

Lesson 1 ①

Key questions

How can we enlarge or reduce this?
What scale is this?
What is the distance?

Vocabulary

Scale, enlargement, reduction, graph.

Introduction 15min

Assemble the scale models you have collected and use them to get the class to suggest how they relate to the real object. As these are three-dimensional objects, the children should demonstrate that they can visualise a reduction of the same order in dimensions in all directions. Get individuals out to measure the length and height of the models and ask the class to estimate by how much the model has been reduced. For example, a large model car may be a 1:32 model of the real thing; thus a 12cm car would be 32 × 12cm long if it was full size.

Activities 25min

Give out geoboards and elastic bands to small groups. Ask them to make a square and then another square twice the size. Can they also make one 3 times as large? What about starting with the largest square or rectangle they can and reducing it by a half and by a quarter? Give out graph paper and talk about the need to be able to convert lengths to a given scale. Tell them that you want a line graph that helps us to convert from centimetres to metres for a plan in which 1cm stands for 3m. Point out the need to have the graph set out with axes like those below. Get the children to draw the line graph.

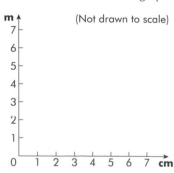

Closing the lesson |15 min|

▓ Bring the children back together and get them to give you some results from their graphs. Choose lengths such as 2.5cm to get a metre equivalent. Work from metres to centimetres too.

Assessment

Child's performance	Teacher action
Can produce geoboard work but finds the graph really quite demanding	Consider doing more work on the construction of straight-line graphs, in particular conversion graphs
Completes all aspects with some help	Revise knowledge about straight-line graphs and their construction
Completes all aspects	Move on to next lesson

Lesson 2   ②

Key questions

How can we obtain an accurate measure?
What scale should we use?

Vocabulary

Scale, plan, measuring vocabulary.

Introduction |15 min|

▓ Give out or display a simple map of the school, with any grounds as well as the buildings. With the children's help, discuss how the plan relates to their real-life experience. Remind them about the scale of the map. Tell the class that, in groups, they are going to create a scale plan of their classroom. Using an appropriate measuring tool, measure the length and breadth of the room and the diagonals. Discuss a scale that would be sensible in drawing a plan of the room.

Activities |30 min|

�ें Using squared paper and drawing instruments (pairs of compasses can be a great help) get each group to produce a floor plan for the room. Then, using available measuring tapes and rulers, they should attempt to locate the main components of the room on their plan. You might choose to use card cut-outs to scale for movable furniture.

Closing the lesson |10 min|

▓ Talk over what has been achieved. It may be necessary to allocate further time for completion of the scale plans that could then form part of a display on scale, maps and related ideas.

Assessment

Child's performance	Teacher action
Produces a floor plan but finds the detail more difficult	Give more time but talk through the processes of measurement and scaling once again
Completes floor plan and some of the components	Give some more time and then move on to next lesson
Completes floor plan and most of the components	Give some more time and then move on to next lesson

Lesson 3   ③

Key questions

What is the scale?
What does this symbol mean?
How can we make good estimates of the distances?

Vocabulary

Scale, measure, maps, plans.

Introduction |15 min|

▓ Give out copies of maps, at least 1 copy between each pair. You might use geography texts or walking books with maps or reproduce, with permission, parts of the appropriate OS map. Get the children to identify key elements such as roads, paths, woods and rivers. Talk about the scale of the map.

Activities |30 min|

◍ The challenge is to identify an appropriate route for a circular walk. The children should consider such factors as how to get to the start, making the route interesting and how strenuous their walk would be. They have to work out how far their walk is and make an estimate of how long it might take. They can make notes on **Copymaster 39**.

Closing the lesson |10 min|

▓ Ask someone in each group to outline their walk; how far it is and why they chose that route.

Assessment

Child's performance	Teacher action
Contributes to the task but finds it hard to relate the map to real distances	Consider further work on ratio and scale
Active in the planning and makes a reasonable estimate of length of time and route	The learning targets for this theme have been met
Well thought out route with a good rationale	The learning targets for this theme have been met

Homework

Get the children to have a go at producing a scale plan of their bedroom, garden or kitchen. Give practice on the production of straight-line graphs. These might be conversion graphs such as kilometres and miles or 2 different currencies.

Investigations

- Do more work on multiplication tables. Try multiplying by 9 and dividing by 9 and looking for patterns. Extend to the 11× table and then try multiplying numbers by 11 to see what patterns can be detected. Also, try putting multiplication table products into a circle and see what patterns can be drawn. The diagram below shows what is meant for one of these.

6× table: final digit of the products

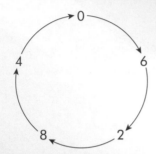

- Explore odds and evens. What happens if we add 2 odd numbers? What happens if we take away 2 odd numbers? What about even numbers? What about odd and even? What happens when we multiply and divide odd and even numbers? Is there any pattern?
- Extend the work on Fibonacci and the Golden Ratio. For example, if you measure the distance from the tip of the finger to the first joint and then to the second and then to the third, is this in a ratio of about 1.62:1? Make a collection of postcards of famous paintings and get the children to measure on these to see if any of the parts of the paintings are in the Golden Ratio.
- Take any three-digit number where the digits are all the same, e.g. 444. Total 4 + 4 + 4, the result of the 3 digits. This comes out as 12. Divide your original, in this case 444, by 12 and you will get the answer 37. Try this for other combinations of three-digit numbers, where the digits are all the same. Can you explain the results?

- Using calculators, key in the number 1089, then multiply it by 1; then 1089 × 2; 1089 × 3 and so on. Set out the answers carefully in columns. What do you notice about the patterns? Try adding the products of each of these multiplication tables, getting down to the reduced number. So, e.g., 1089 × 2 = 2178. If we add 2 + 1 + 7 + 8, we get 18. If we add 1 + 8 we get 9. Does this work in the same way for all the products?
- Write down 3 different numbers less than 10, e.g. 2, 4, 7. Make all possible two-digit numbers from these. There will be 6. In this case, they are 24, 27, 74, 72, 42 and 47. Total these. They sum to 286. Now go back to the original 3 numbers, 2, 4, 7. Add these together, 2 + 4 + 7 = 13. Divide the total of 286 by 13. What do you get? Now try it for 3 different numbers less than 10. What do you notice?
- Explore Napier's bones or rods. The diagram below shows you Napier's bones. How do they work?

Make the rods from lollystick or card strips.

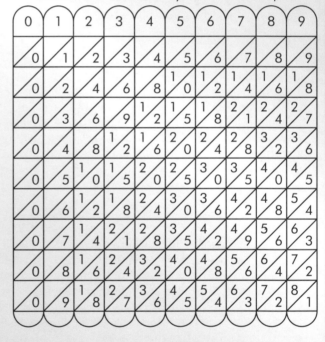

Assessment

- Regular multiplication table practice.
- Periodically test tables knowledge. In doing so, bear in mind the work we have done on the 'hardest' multiplications. Also give questions which give the commutative nature of multiplication, e.g. 3 × 6 and 6 × 3. Additionally, see if the children can come up with different multiplications that produce the same products, e.g. 9 × 4 = 36 and 6 × 6 = 36.

- Test children's ability to multiply by multiples of 10. Include 10×, 20×, 50× and 100×.
- In support of their understanding of division, assess the children regularly on their ability to do subtraction quickly and accurately.
- Give problems in which decisions need to be taken about remainders. For example, how many rolls of wallpaper at 57p can I buy for £5.00?

MENTAL ARITH
AND NUMBER PA

The advent of the calculator has placed even greater emphasis on the use of mental methods for estimation, checking and efficiency. The ability to add 2 two-digit numbers is undoubtedly an important one. To do this there is a whole range of known facts that children have to acquire. They then have to be able to use those known facts to derive new information. So, e.g., being able to do, in one's head, $26 + 7$ by knowing that $6 + 7$ is 13 and then knowing that we can add the 20 to make 33 is crucial. Similarly, to be able to handle $55 - 17$ mentally the children need to be shown a variety of ways of tackling a problem. We should not simply expect a child to 'catch' the way to do things. As has been the case with times tables, we need to overtly teach a range of techniques and strategies, and provide well-structured opportunities for children to expand their bank of known facts in addition and subtraction.

The starting point for mental mathematics is counting on. In learning the counting numbers children are encouraged to see this as an appropriate way of thinking about numbers. Counting on in 2s, 5s and 10s is a common experience – or should be – for youngsters. In work on their times tables we can use the fact that multiplication can be viewed as repeated addition and reinforce the counting on strategy. Such a strategy helps with problems like $46 + 27$, where a child can count on the tens 40, 50, 60 and then add the 13 ($6 + 7$).

However, whilst counting on is a powerful tool it has dangers. Any child who sees counting on in 1s or 2s as the only available way of solving mental problems is doomed to frustrated failure. Imagine, e.g., trying to solve $37 + 14$ by counting on in 1s. Especially if you are not allowed to use your fingers. Children need to know that they can move numbers around, crack them open, and recombine them providing that they abide by the rules. So the use of commutativity (7×8 may seem easier than 8×7, and $8 + 2$ feels easier than $2 + 8$) is important. To be able to use the associative and distributive laws too will make a huge difference to the ability to handle mental problems. And whilst doing the written algorithm in one's head is, in some circumstances, a possible strategy, an over-emphasis on such algorithms inhibits the flexibility needed to handle a variety of mental problems.

Whilst a variety of examples have been given here and in the rest of the section, it cannot be overstated that children's personal methods – if accurate – need to be encouraged. There is a variety of ways of doing mathematics in your head and no one way is more 'right' than any other, providing the results are correct. The most important question in supporting children's enthusiastic development of sound mental arithmetic must be: How did you do that?

Addition and subtraction

Learning targets

On completion of this theme the children should be able to:

1 ➡ demonstrate a range of known facts in addition and subtraction
2 ➡ order and partition given mental problems
3 ➡ use known facts to derive some new facts.

Before you start

Subject knowledge

There is now clear evidence that those children who develop a variety of strategies for tackling mental arithmetic problems are more likely to succeed than those who use only the early strategies such as counting on. It is important, therefore, to draw to children's attention the fact that we can actually play with number problems, using a few simple rules. To develop speed and accuracy children need to make use of the facts that they know to derive new facts. Faced with an addition like $7 + 17$ the more confident children might reverse this as we know that order can help and that taking the larger number first usually simplifies the problem. They might also use the known fact that $7 + 7$ equals 14 (two 7s), and then we need to add 10. This involves the child in partitioning the 17 into $10 + 7$. Such an approach genuinely helps. The rule that the children need to know in this example is that of associativity. This example illustrates two possible

uses of the associative law for addition:

$$15 + 7 = (10 + 5) + 7 = 10 + (5 + 7)$$
$$= 10 + 12 = 22,$$

or

$$15 + 7 = 15 + (5 + 2) = (15 + 5) + 2$$
$$= 20 + 2 = 22.$$

Previous knowledge required

Lots of experience with common number bonds in addition and subtraction.

Resources needed for Lesson 1

Copymaster 40.

Resources needed for Lesson 2

No special resources required.

Resources needed for Lesson 3

No special resources required.

Teaching the lessons

Lesson 1 ❶

Key questions

What is the sum of these numbers?
What is the difference between these numbers?
Can you see a pattern?
How did you work that out?

Vocabulary

Addition, subtraction, order, pattern.

Introduction 10min

▨ Start the lesson with a brisk set of questions centred on the common number bonds that the children should already know. Keep these questions to single-digit pairs of numbers at this stage. When the class is nicely warmed up and attuned give out **Copymaster 40**.

Activities 20min

👤 Each child should complete the addition squares on

the copymaster. If there is time some may go on to create some of their own. Encourage the children to look for any patterns they can see in their squares.

Closing the lesson 10min

▨ Go over some of the questions the children have been tackling, with their help. Pick up patterns where appropriate. Finally, try a few common subtraction problems that are the inverse of the addition problems they have been working on.

Assessment

Child's performance	Teacher action
Has some of the common number bonds but is inclined to be trying to count on using fingers	More practice is needed in memorising the common number bonds – may mean the use of practical apparatus and discussion rather than a rote learning approach

64

Copes satisfactorily with the core activity — Move on to next lesson

Completes the activities and identifies clear patterns within answers — Move on to next lesson

Lesson 2 ②

Key questions

How did you do this calculation?
What other ways might you do this?
Has anybody another way of doing this problem?

Vocabulary

Addition, order, partitioning (or 'splitting' or 'cracking open') the number.

Introduction ⌈15min⌉

Warm up the class with some practice in common addition number bonds. Then put a couple of examples of additions on the board, e.g. 12 + 8 and 6 + 9. Whilst these may be known addition facts for most of the children, you can use them in order to illustrate a particular strategy. Split up the first sum like this:

12 + 8 = 10 + 2 + 8

and ask the children why this might be an interesting approach to the problem. Reorder the sum to 10 + 8 + 2 in this discussion. Now do something similar with the second sum along these lines:

6 + 9 = 5 + 1 + 9 = 9 + 1 + 5 = 10 + 5

and again engage the children in discussion. If they are happy with this then move on, if not work through some more examples

Activities ⌈20min⌉

Put a series of addition problems on the board. These should include two-digit plus one-digit, as well as one-digit only problems. Depending on the class you might also include some two-digit plus two-digit problems. Get the children to work with a partner to come up with at least 2 ways of organising the problems as well as offering a solution to each problem.

Closing the lesson ⌈10min⌉

Use examples from the work of pairs of children to go over what they have been learning and to reinforce the approach.

Assessment

Child's performance	Teacher action
Can see 1 approach only	This may require further explanation and in some cases further work with mathematical apparatus
Provides 2 approaches and obtains the solution correctly	Move on to the next lesson checking whether child is using this sort of strategy in other contexts
Does the activity readily and offers further useful suggestions	Move on to next lesson

Lesson 3 ③

Key questions

How did you solve this?
When is it a good idea to split up the numbers?

Vocabulary

Addition, order, partitioning (or use 'splitting' or 'cracking open') the number.

Introduction ⌈15min⌉

This lesson builds closely on Lesson 2. Again warm the class up with number bond practice, including some subtraction this time. Put up on the board similar problems to those in the last lesson but include some of this type: 19 + 5, 16 + 9, 31 − 9 and 23 − 8. Remind the children of the idea of splitting up and reorganising the problems to simplify them. Work through your examples on the board paying particular attention to the ways in which partitioning can help with subtraction.

Activities ⌈20min⌉

Working collaboratively the children have to make up some problems in which the splitting up of numbers is helpful. The problems they identify will be collected to form the basis for a class booklet of problems.

Closing the lesson ⌈15min⌉

Ask pairs of children to give examples of what they have come up with and invite the rest of the class to solve these problems. Collect in what has been done as a resource for follow-up work.

Assessment

Child's performance	Teacher action
Copes with the addition but not with subtraction	More practice needed and possibly extra help and work on subtraction
Works well and comes up with some useful ideas	The learning targets for this theme have been met
Makes a variety of pertinent suggestions	The learning targets for this theme have been met

Homework

Consider giving some more examples for practice at home. Invite the children to explain to parents the methods they have been learning. Offer the children the challenge of producing a track game which requires the solving of mental arithmetic problems.

Developing strategies

Learning targets

On completion of this theme the children should be able to:

1 ➡ engage in mental addition of more than 2 numbers
2 ➡ adjust and readjust numbers in order to simplify problems
3 ➡ use doubles and doubling in their mental arithmetic work.

Before you start

Subject knowledge

A key strategy in successful, speedy and accurate mental arithmetic is the use of the nearest 10. For example, in adding 19 and 23 it is easier to round the 19 to 20 and add this to the 23, and then take away 1 at the end. Such mental manipulation of numbers seems to come to some children 'naturally' but for many we need to clearly, and overtly, teach some of the important strategies. This theme is concerned with just such teaching. In addition to adjusting and readjusting numbers and solutions there is also evidence that the use of doubles is significant in successful mental manipulation.

There is a long tradition of doubling and halving in mathematics. Whilst most children will know their

2× table readily, and be able to double many numbers, it does not always occur to them that this knowledge can be employed to solve a range of mental arithmetic problems.

Previous knowledge required

Multiplication tables, common number bonds in addition and subtraction, rounding to the nearest 10.

Resources needed for Lesson 1

Copymaster 41.

Resources needed for Lesson 2

No special resources required.

Resources needed for Lesson 3

Copymaster 42, Copymaster 43, scissors, counters.

Teaching the lessons

Lesson 1 ①

Key questions

What is this to the nearest 10?
How did you solve this one?
Which facts did you know which helped you?

Vocabulary

Addition, subtraction, rounding, to the nearest, adjust.

Introduction 15min

Warm up the class with a round or two of known fact challenges involving 2 single-digit numbers. Then ask some questions which involve the addition of 3 single-digit numbers. When ready, throw in some which require both addition and subtraction with 3 single-digit numbers, e.g. 3 + 4 − 2.

Activities 25min

Give out **Copymaster 41**. The children have to work through the copymaster giving the ways in which they achieved their solutions and trying to come up with alternative ways. Whilst the material has been structured

to encourage the use of partitioning the numbers so that, effectively, the children are manipulating more than 2 numbers, do accept appropriate alternative strategies – including 'I just knew it.'

Closing the lesson 10min

Choose, with the help of the children, some of the problems from the copymaster and talk them through with the class. Invite a whole range of examples of strategies for their solutions.

Assessment

Child's performance	Teacher action
Relies very much on counting on	More basic work is still needed on common number bonds
Can manipulate some of the problems in a variety of appropriate ways	Move on to next lesson but monitor progress
Works through providing a range of alternatives including many known facts	Move on to next lesson

Lesson 2

Key questions

What is this number to the nearest 10?
What other ways are there?

Vocabulary

Addition, subtraction, rounding, to the nearest, adjust.

Introduction `20 min`

Write up some numbers such as 21, 9, 32 and 19 and ask what these would be to the nearest 10. Now put an example of a problem on the board that involves the addition of a two-digit and a single-digit number, e.g. 19 + 5. Suggest to the children that we would find this easier if it was 20 + 5. Talk through the fact that we can adjust the 19 to 20 (the nearest 10) and subtract 1 from 25 to get the answer to 19 + 5. Use further examples of this sort if necessary to get this message across. Write up some more examples including: 21 + 13, 9 + 17, 8 + 19, 45 + 9, 19 + 9, and 9 + 22. Get the children to attempt these mentally and elicit answers around the class.

Activities `20 min`

Get the children to write down the problems you have just been using and then, collaboratively in small groups, to write down as many ways of doing these as they can think of. Remind them of rounding to the nearest 10 and then adding or subtracting at the end of the process to adjust for the correct solution.

Closing the lesson `10 min`

Get some of the children to put their strategies for solving the problems on the board and explain them to the others. Do a show of hands to see how many others had found the same approach. Reinforce the idea that there are a variety of ways of solving problems mentally and none of them is necessarily right or wrong – unless, of course, they produce incorrect answers.

Assessment

Child's performance	Teacher action
Comes up with a limited range of strategies including counting on	Give further practice on the range of strategies you have been covering in this and earlier lessons
Provides a good range of approaches	Move on to next lesson
Has a great variety of ways of solving the given problems	Move on to next lesson and ask for a contribution to the class database on 'good' problems

Lesson 3

Key questions

What is twice this?
What is double/half this number?

Vocabulary

Double, doubling, half, halving.

Introduction `15 min`

Work briskly through the 2× table. Then give a number and ask what is double this number? And double the answer? And double that? This will be quite demanding but start with 1, 2, 5 or 10 where the children will already know the pattern of numbers. See how far the class can get and return to this exercise subsequently to see if they can create a new 'record'. Now ask some addition questions like 8 + 8, 3 + 3, and 7 + 7. Are the children doing a multiplication or an addition? Move to multiples of 10 and try doubling 10, 20, 50, and 100.

Activities `30 min`

Give each pair of children **Copymaster 42** and **Copymaster 43**. They need 20 counters. The children have to play Bingo. They cut up their Copymaster 42 to make 'cards' which are then shuffled and put in a pile in the centre of the table. Each player takes a 'card' and sees whether they think they can cover a number in one of the tables on Copymaster 43 with a counter. All 'cards', including those for which there is no match, go to the bottom of the pile each time. The winner is the one who covers all the numbers in a table first.

Closing the lesson `5 min`

Ask the children which of the cards they found easiest to solve and which the hardest – and why this might be the case.

Assessment

Child's performance	Teacher action
Has some difficulty in moving at the speed partner wants	Further practice of multiplication tables and common number bonds is necessary
Plays competently and is satisfied with own performance	The learning targets for this theme have been met
Is very rapid, confident and accurate	The learning targets for this theme have been met

Homework

Make more copies of Copymasters 42 and 43 for the children to take home and play with family and friends.

Missing numbers

Learning targets

On completion of this theme the children should be able to:

1 ➤➤ identify the missing numbers in a problem
2 ➤➤ work with simple function machines
3 ➤➤ work with function machines with mixed operations.

Before you start

Subject knowledge

Mathematics is essentially about pattern. On the basis of pattern we can go on to make predictions and general statements. In developing ideas about what will make a process work, children are laying foundations for good comprehension of the mathematics they will encounter both later in education and in everyday life. Using letters or symbols is a succinct way of making a mathematical statement but, in the early stages, it is easy to lead children to some erroneous beliefs – the main one being that we need always to get an answer which is a number. For example, in physics we can state that a force is the product of mass and acceleration and write this as $f = ma$, and as a general statement we need not go any further. Only if we want to calculate a particular force do we need to have measures of mass and acceleration. So children need to be encouraged to see the use of letters and symbols as an efficient way of making a statement. In working with function machines it is crucial that the children appreciate that the function is fixed. We can vary the inputs to a function machine but, regardless of the input, the machine will operate on it in the same way each time.

Previous knowledge required

The four operations.

Resources needed for Lesson 1

No special resources required.

Resources needed for Lesson 2

Copymaster 44.

Resources needed for Lesson 3

No special resources required.

Teaching the lessons

Lesson 1   ❶

Key questions

What number am I thinking of?

Vocabulary

Symbol.

Introduction |15min|

▓ Play a guess the number game of the sort: 'I am thinking of a number. If I add 3 to it I get 8. What is my number?' Do this again using multiplication, then subtraction and then division. Invite individuals to have a go at challenging the class. Show them, on the board, that you could write $\square + 3 = 8$ for your first example. Then show them that you could have the 'missing number' in any position, e.g. $6 \times \square = 12$.

Activities |20min|

👥 Each pair of children has to produce at least 5 examples of missing number statements. They should use the symbol $\square$ in different places, and they should be encouraged to have examples that use different operations.

Closing the lesson |10min|

▓ Get the class back together and nominate some of the pairs to give 1 of their statements for the class to solve. Collect in the work with a view to using the examples in a follow-up lesson.

Assessment

Child's performance	Teacher action
Produces a limited range of statements	Give further practice, using exercises from appropriate scheme books or other texts
Completes the task but is not confident with all operations	Give further practice at generating statements
Completes the task using all combinations	Move on to next lesson

Lesson 2   ❷

Key questions

What does the function machine do?

Vocabulary

Function, function machine.

Introduction

Introduction

`15 min`

⊞ Draw the function machines shown below on the board, one at a time.

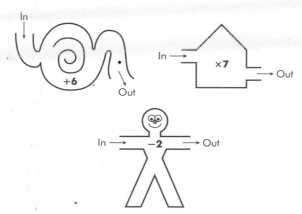

Work through each example explaining to the children what they are called and what they do. Get the children to suggest input and you give the output. Then ask some to give both input and output. When you feel they are prepared give out **Copymaster 44**.

Activities

`20 min`

👤 The children should work through the copymaster indicating output or input then attempt to create some function machines of their own.

Closing the lesson

`10 min`

⊞ Go over some of the examples from the copymaster so that the children can self-assess. Then ask for some of their ideas for new machines.

Assessment

Child's performance	Teacher action
Completes most of the given examples but finds it hard to generate their own.	Give more examples to work through then, with close support, attempt to get child to create own once more
Completes the copymaster making some suggestions of their own	Move on to next lesson
Completes the copymaster and comes up with a range of ideas possibly including putting an input through more than one machine	Move on to next lesson but consider giving some extra time to see if child can make suggestions work

---

Lesson 3 ③

Key questions

What will be the output/input here?
What if we connected two machines?

Vocabulary

Function, function machine.

Introduction

`15 min`

⊞ Remind the children of the work they have already done. Tell them that in this lesson they are going to be tackling challenges where the machines will be doing 2 things to each input. Work through some examples like those below.

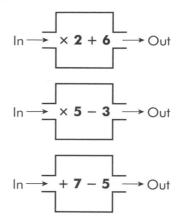

Tell the children that they are going to work in pairs to generate some function machines of their own which must have at least 2 operations in them. Bear in mind that this could generate such things as negative numbers so be prepared to guide those who might get stuck.

Activities

`25 min`

👥 Working in pairs the children should invent some function machines within the constraints that you have indicated. They should make a neat copy of at least one machine that other children could attempt later. As you go round the class choose some machines to share with the class at the end of the lesson.

Closing the lesson

`10 min`

⊞ Share the ideas you have picked up during the activity. Finish by putting a machine on the board that has a range of inputs and associated outputs and ask the children to suggest what the function of your machine might be. There could be several possibilities, all of which work.

Assessment

Child's performance	Teacher action
Helps produce some examples but makes obvious mistakes	Consider rehearsing known common number bonds and building these into simple sequences
Seems to have a satisfactory grasp of the idea though makes a few errors	Further practice may be needed when these ideas are revisited
Completes the task effectively	The learning targets for this theme have been met

Homework

Produce some sheets with various drawings of function machines and invite the children to use them to produce challenges that they would like to offer to the class. Instigate a search for 'Think of a number …' games and ask the children to find examples in puzzle books and reference books.

Money

Learning targets

On completion of this theme the children should be able to:

1 ➡➤ carry through an investigation using coins

2 ➡➤ explore patterns in using combinations of coins

3 ➡➤ make choices and evaluate the costs of these.

Before you start

Subject knowledge

To investigate and extend knowledge of money at this stage it is important to involve the children in a variety of investigations, role-playing games, and manipulation of coins and notes to accumulate total amounts. In so doing, not only are the children gaining further insights into money and money matters, but have a wide range of opportunities to rehearse and reinforce basic mathematical skills and ideas. These include the use of common number bonds, and practice with the four operations. There are opportunities for estimation, pattern identification, and experience in rounding up and down. There is also an opportunity to keep reminding children that, for example, we cannot write £1.5 but must either write £1.50 or £1.05 – and the difference between these is significant.

Previous knowledge required

Four operations, experience with money, knowledge of coins and notes, and their relationship.

Resources needed for Lesson 1

None special though you might consider having imitation coins available.

Resources needed for Lesson 2

None special though you might consider having imitation coins available.

Resources needed for Lesson 3

Copymaster 45, catalogues with goods and their prices – as many as possible and of as many types as possible (these can be dated but not too old), calculators.

Teaching the lessons

Lesson 1 ①

Key questions

How many coins do you need to make this total? How many ways can you do this?

Vocabulary

Money names and associated language, total, sum, amount.

Introduction `10 min`

 Quickly write up, following the children's instructions, all the coins and notes we have up to and including £20. Ask some questions about how the coins and notes relate one to the other. For example, ask how many 10p coins in £1 and how many 2p coins in 50p and so on. Use this as an opportunity to cover some mental arithmetic that should be within the ambit of most of the children. Then tell the class that, working in pairs, they are going to do an investigation involving 2p and 5p coins.

Activities `20 min`

 Ask the children to imagine that they have a supply of 2p and 5p coins (if you feel it appropriate use imitation coins). Then pose the question that they are to investigate: 'Can you make all totals from 2p

to 20p inclusive just using 2p and 5p coins? Go on to ask whether there are different ways of making these amounts, and any amounts we cannot make. Invite the children to come up with the ways they can make each amount, trying to find the least number of 2p and 5p coins in each case.

Closing the lesson `10 min`

Ask which amounts could not be made. If the children disagree get pairs to share their solutions. Choose a few examples where there are different ways of making the total amount – these include 12p, 16p and 18p. How have the children resolved these?

Assessment

Child's performance	Teacher action
Has some difficulty with the mental arithmetic challenges	Further practice needed especially in relation to multiplication tables, multiplication and division
Produces a good response to the investigation but cannot find all solutions	Move on to next lesson but find time to guide child through parts of the lesson again
Identifies the solutions correctly and identifies those that cannot be done	Move on to next lesson

Lesson 2

Key questions

What did you find?
Is there a pattern?

Vocabulary

Money vocabulary, investigation, total, sum, amount.

Introduction `15 min`

 Write up on the board the amount 6p. Ask the children to tell you in how many ways this amount could be made using 1p, 2p and 5p coins. You are looking for 5 ways: 6 × 1p; 4 × 1p and a 2p; 2 × 1p and 2 × 2p; 3 × 2p; and a 5p and a 1p. Tell the children that they are going to attempt this exercise to find out how many ways they can make 11p.

Activities `25 min`

 Working in pairs and using 1p, 2p, 5p and 10p coins get the children to work out in how many ways they can make a total of 11p. If there is time get the children to try a total of their own choosing, but no greater than 15p.

Closing the lesson `10 min`

 Put the combinations that the children have found for the original 11p problem on the board. There are, in fact, 12 different combinations. Discuss how we might organise ourselves to tackle such an investigation. How did the children organise their work, and how do we make sure we have not missed any combinations?

Assessment

Child's performance	Teacher action
Finds some of the combinations but fails to organise work	Work through some smaller totals indicating ways in which we might set out our work logically
Finds all, or can readily see in the closing session how to find all, possible combinations	Move on to next lesson
Finds all combinations and attempts another amount of own choosing	Move on to next lesson but invite child to pursue this further at another time

Lesson 3

Key questions

What are you aiming to achieve?
How did you make your choices?

Vocabulary

Investigate, calculate, total, amount, sum, product.

Introduction `10 min`

 The aim of this session is for the children to collaborate in itemising and costing a particular sort of event. The lesson may need follow-up time – it could form part of a topic. You may choose to put a limit on what they can 'spend'. Choose viable tasks that you can resource. For example, the children could be kitting themselves out for a camping expedition, or catering for a visiting class from another school, or organising the purchase of gifts and party materials for a celebration. Depending on the type and range of catalogues you have managed to accumulate either set the children the same sort of task, or disperse different catalogues and tasks across the groups.

Activities `30 min`

 If appropriate **Copymaster 45** can be cut up and the tasks distributed amongst the class. The groups work at the allocated task. Go round asking them about their choices and reminding them of forgotten items.

Closing the lesson `15 min`

 Get each group to report back on what they 'spent' and some of what they chose and why.

Assessment

Child's performance	Teacher action
Participates but finds the costing difficult	May need some more formal practice at money problems
Participates and has no real problem with costing	The learning targets for this theme have been met
Participates and evidences a clear appreciation of the need to be accurate and balance the books	The learning targets for this theme have been met

Homework

The investigations with coins could easily be extended and the children could explore combinations of coins with a variety of coins available. The final lesson could provide a wide range of homework possibilities specific to the class and the tasks you chose.

Learning targets

On completion of this theme the children should be able to:

1➡→ determine factors for given numbers and understand their connection with multiples

2➡→ identify prime numbers

3➡→ explore prime factors.

Before you start

Subject knowledge

Through their work on multiplication, including the learning of tables, the children should have come to the appreciation and an understanding of the commutative law, e.g. that $3 \times 6 = 6 \times 3$. Indeed they should understand that, in this case, 6 and 3 are factors of 18. Of course 18 has other factors, the set being 1, 2, 3, 6, 9, and 18. The children should also appreciate that, e.g., 5 is not a factor of 18. Finding the factors of a natural number is an important step in understanding both that number and its relations to others but is also a fundamental idea in multiplication, division and the manipulation of fractions. In looking at prime factors the children are taking on another important idea. Every number is divisible by 1 and itself – in some cases these are the only factors a number has. These prime numbers are important for a variety of reasons. Early mathematicians worked on primes and Eratosthenes, a Greek who lived in the period around 250 BC, developed a method for identifying prime numbers which we call the Sieve of Eratosthenes. This method is explained in Lesson 2.

Prime numbers continue to be the source of much mathematical investigation. Here the key idea is that there are ways of finding prime numbers, and that all other natural numbers can be made from the multiplication of prime numbers. This last idea has been met before in the suggestions made for teaching and learning multiplication tables.

Previous knowledge required

Multiplication, multiplication tables, division.

Resources needed for Lesson 1

Copymaster 60.

Resources needed for Lesson 2

An overhead transparency of a 100 square (this could be made from Copymaster 46) – failing this put the first three rows of a hundred square on the board, Copymaster 46, coloured pencils.

Resources needed for Lesson 3

No special resources required.

Teaching the lessons

Lesson 1 ①

Key questions

What are some of the multiples of this number?
What are the factors of this number?
How many factors does this number have?

Vocabulary

Multiple, factor, prime number.

Introduction | 15 min |

Write a number such as 12, 18 or 24 on the board. Ask the children how many ways they can tell you of making the chosen number by multiplying whole numbers. List these on the board until all have been found. List the multiples in a row from smallest to largest. Point out that there are no more and that the selected number can be divided by any one of the multiples, and the answer will be one of the other multiples. Explain that we call the set of multiples the *factors* of the number. If necessary do the exercise again with another number, but avoid prime numbers at this stage.

Activities | 25 min |

👤 Put up a set of numbers, both odd and even but not prime, on the board and ask the children to identify their factors. Using the squared paper the children can draw rectangles to help determine all of the factors. See below for the rectangles for 15.

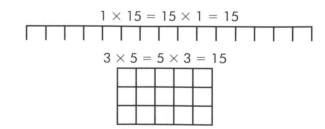

$$1 \times 15 = 15 \times 1 = 15$$

$$3 \times 5 = 5 \times 3 = 15$$

Closing the lesson | 10 min |

🎲 Go over one or two of the numbers you gave as a self-assessment aid for the children. Now put up on the board 1 or 2 prime numbers and with the children determine the factors. Tell them that these are known as *prime numbers* if they have not met the term before.

Assessment

Child's performance	Teacher action
Has problems in developing all of the factors for particular numbers	Check on knowledge of multiplication tables then return to this exercise
Completes the exercise with some help	Move on to next lesson but consider giving further opportunities to practise finding factors
Completes the exercise	Move on to next lesson

Lesson 2 ②

Key questions

What makes a number prime?

Vocabulary

Multiplication, division, multiple, prime.

Introduction

 Using the overhead transparency or the board tell the children that they are going to find all of the prime numbers up to 100 using a method named after an ancient Greek mathematician, Eratosthenes. Identify the first few yourself and explain that you are shading all of the multiples of 2, 3 and so on because they cannot be prime.

Activities 20 min

 Give out **Copymaster 46**. The children need to work through the copymaster colouring those numbers that are composite and leaving the prime numbers unshaded. Help with the first row particularly.

Closing the lesson 10 min

 With the help of the children write up the prime numbers up to 100. Draw their attention to the fact that there seem to be fewer primes as we go along, and that some primes seem to come in pairs either side of a composite number.

Assessment

Child's performance	Teacher action
Makes progress but does not complete the Sieve of Eratosthenes	Give more time but check child is clear about the rules needed to operate
Completes the copymaster with some help	Move on to next lesson
Completes the copymaster with little or no help	Move on to next lesson

Lesson 3

Key questions

What are the factors of this number?
Which of these factors is prime?
Can you see a way of making the number using just the prime factors?

Vocabulary

Multiple, factor, prime.

Introduction 10 min

 Start by putting 3 single-digit prime numbers on the board. Ask what composite numbers can be obtained by multiplying these numbers. An example appears below.

$$2, 3, 5$$
$$2 \times 3 = 6, \ 2 \times 5 = 10, \ 3 \times 5 = 15$$
$$2 \times 3 \times 5 = 30$$

Activities 25 min

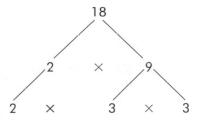 Put up another 3 primes and ask the children to work together in pairs to find all the composite numbers.
Draw the children back together and show them a way of developing the multiplication of primes starting with a given composite number, e.g.:

They may have met this approach before in the work done earlier in this section on multiplication tables. However, give them further practice, working in pairs. Put up some numbers for the children to start with to make their tree diagrams.

Closing the lesson 10 min

 Revise what the children have so far learned about factors, multiples and prime numbers.

Assessment

Child's performance	Teacher action
Accomplishes first part of the activity	Give more time and/or use some of the multiplication tables ideas
Tackles both parts with help	Assess understanding of factors and then prime factors
Works through with little difficulty	The learning targets for this theme have been met

Homework

Invite the children to explore numbers that are important to them. For example, for each child, where are the prime numbers and what are the factors of the numbers in: his or her birthday; his or her age; his or her lucky number; his or her house number; numbers in the family car registration; bus number caught to school; number of the house opposite?

73

Exploring patterns

Learning targets

On completion of this theme the children should be able to:

1 ➡➡ solve a simple magic square

2 ➡➡ discuss some attributes of square numbers

3 ➡➡ see why triangle numbers are so called and investigate some ideas about such numbers.

Before you start

Subject knowledge

The ancient Greeks were interested in shape and space, and this interest is reflected in many of the names for patterns of numbers. Square numbers and triangle numbers are 2 examples of the way in which the Greek mathematicians of old have left us a legacy of the link between shape and number. Square numbers are significant in 2 main settings – multiplication of numbers and early work on powers. Triangle numbers arise in many circumstances in mathematics and an introduction here will be useful in subsequent work. Magic squares are also ancient. There is a 3 × 3 magic square in a Chinese text which dates from 1000 BC.

There are many strategies for approaching magic squares. Odd number squares are easier than even number squares (try a 4 × 4 to see this). The key requirement is that all columns, rows and diagonals must sum to the same total – the centre square in odd number magic squares has a relationship with this total. One solution for a 3 × 3 magic square is shown here. If you add the same number to all the numbers in the square, or subtract the same amounts, or multiply or divide by the same number the magic square still works.

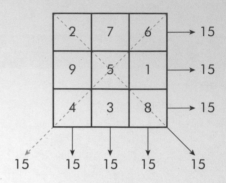

Previous knowledge required

The four operations, multiplication tables.

Resources needed for Lesson 1

Copymaster 47.

Resources needed for Lesson 2

Copymaster 60 or other squared paper with squares of a different size, some square plastic templates if needed, coloured pencils.

Resources needed for Lesson 3

Counters, and triangular dotty paper found on Copymaster 70 may be useful.

Teaching the lessons

Lesson 1 ①

Key questions

Do the rows/columns/diagonals sum to the same total?
Do you notice anything about the number in the centre square?

Vocabulary

Row, column, diagonal, sum, total, magic square.

Introduction `10 min`

▨ Tell the children that they have a challenge to meet today that is easy to explain but may take a little while to resolve. Draw a 3 × 3 matrix on the board and write the numbers 1 to 9 inclusive below it. The

challenge is to put these numbers into the cells of the square using each one once only. The catch is that the rows, columns and diagonals must all sum to the same total. There are alternative solutions.

Activities `25 min`

▨▨ The children should attempt to construct a 3 × 3 magic square using **Copymaster 47**. When solutions start to appear ask the solvers, after checking their solution meets the criteria, what happens if they add, say 8, to each number in their square. Then take away 5, or multiply by 3, or divide by 2?

Closing the lesson `15 min`

▨ Put up one of the solutions to the 3 × 3 magic square. Draw the children's attention to the centre square and ask what relationship this has with the totals for rows, etc. Ask those who got that far, what happens if you add and so on. Tell them that other magic squares of different sizes can also be produced.

Assessment

Child's performance	Teacher action
Completes magic square	Move on to next lesson but consider returning to this topic at a later date
Completes magic square and attempts other solutions	Invite child to try manipulating his or her solution using all 4 operations; move on to next lesson
Completes magic square and extends using some or all of the 4 operations	Challenge child to produce a 5 × 5 magic square; move on to next lesson

Lesson 2   ②

Key questions

What is the amount by which the pattern grows each time?
Can you see a connection in the pattern?

Vocabulary

Square numbers.

Introduction   [10 min]

Draw a square on the board, then 4 squares of the same size making a larger square. Talk about the fact that we could call the first 'one squared' and the second 'two squared'. Now add, 'three squared'. Tell the children that you want them to draw these squares for themselves, colouring the squares that are added each time in a different colour. When they have done this for a few more you are going to be interested in what patterns they can see.

Activities   [20 min]

The children should colour from a 1 square progressively to at least a 5 × 5 square onto squared paper. Then ask pairs whether they can see a pattern.

Closing the lesson   [15 min]

Use your original drawings, adding more up to a 5 × 5 square. Ask the children what they notice. If the answer does not come then talk them through the fact that we have added 3, 5, 7, and 9 to get the next largest square. Now point out that 1 + 3 = 4 (two squared), 1 + 3 + 5 = 9 (three squared) and so on.

Assessment

Child's performance	Teacher action
Makes and colours squares but is not confident about a pattern	Take child through the exercise again at a later stage
Sees the pattern of growth	Move on to next lesson
Sees that adding the set of odd numbers produces a square number	Move on to next lesson

Lesson 3

Key questions

How many more dots, or counters, each time?

Vocabulary

Triangle number, triangular.

Introduction   [15 min]

Draw a small circle on the board and then 3 similar circles in the shape of a triangle. Then, after eliciting the shape, draw the next largest triangle of 6 similar circles. As with the work on square numbers the first challenge is for the children to reproduce this sequence and then extend it. Give out counters and, if you wish, triangle dotty paper on **Copymaster 70**.

Activities   [25 min]

The children should reproduce your examples and then extend them making the next largest triangle each time. When they have done a few ask them to work out what is being added each time. Try to get them to see the pattern in 1, 1 + 2, 1 + 2 + 3, 1 + 2 + 3 + 4 and so on.

Closing the lesson   [10 min]

Draw the class back together and quickly recap what they have been doing. Now extend this by pointing out that the series of totals they have, that is 1, 3, 6, 10, 15 and so on, are called the *triangle numbers* – why is this the case? Then, if appropriate, point out that if we add consecutive triangle numbers like 1 + 3 = 4, 3 + 6 = 9, and 6 + 10 = 16 we produce another familiar set of numbers – what are these?

Assessment

Child's performance	Teacher action
Makes the triangles but is not confident about a pattern	More work on this needed but at a later date
Sees the pattern of growth	Seek ways of giving more investigative work in number
Sets out the pattern	The learning targets for this theme have been met

Homework

Concentrate on magic squares as these can generate a lot of interest at home too. For some, give challenges to do with addition and multiplication. With others use subtraction and division also. Where appropriate, offer challenges for larger odd number squares and then, if possible, even number squares.

Investigations

- The children roll 2 dice and add, subtract and multiply the resulting numbers. Add another die and get them to try addition, mentally, of 3 numbers. Then move them on to multiplying the first 2 and adding or subtracting the third. More variations can be built up and the children can suggest these.
- Make a pack of cards with numbers on them. The children have to draw a number and then, using mixed operations, produce that number.
- There are software packages to help with practice in the four operations. Increasingly there are now CD-ROMs that offer some interesting mathematical problem solving.
- The children can research the 4 operations in history and in other cultures. The abacus might be a good starting point. They could look at Roman numerals to see how they might have been added or subtracted.
- Draw graphs of the square numbers. The x-axis has numbers up to, say 100, and the y-axis to 10. What shape is the line? Now try the triangle numbers.
- Magic squares of orders greater than 3 can be attempted. The odd numbers are easier than the even numbers. See if the children can find out more about the Chinese magic square, the lo-shu, the work of Euler, and the fact that Dürer included a magic square in one of his paintings.
- Extend the finding of factors by using calculators, including factors of some numbers up to 100.

Assessment

- Give times table tests both orally and in written form.
- Give written exercises in all operations periodically. These should take the form of both number problems and word problems.
- The children should produce a given number of challenges in particular operations for others to try. Again these should be in number form and written form.
- Use opportunities in measures work, including money, to evaluate the computational skills of the children.
- The children should produce an explanation of a particular operation to help others do that kind of calculation.
- The children have to produce a game that involves the use of at least 2 operations.

HANDLING DATA

Data handling and the extraction of meaning from data is core to our activities as human beings. From birth we are sorting and classifying what we encounter and trying to construct a view of the world that makes sense and allows us to feel secure. In modern life we also face a veritable plethora of information through the media and new technologies, all of which we need to sort in order to manage our complicated lives. Whether it is organising our budgets, paying our taxes, or buying and selling expensive items like houses and cars, we have to be able to use information, charts, tables and probabilities. For these reasons there is no doubt that handling data is an important aspect of mathematical learning for youngsters.

This learning is not confined to mathematics lessons, of course, as many school subjects make use of statistical ideas. In geography we might expect children to understand something about populations, or weather patterns. In history we might want the children to compare and contrast living conditions. In science we ask them to classify plants and animals, and to collect data about the environment. In English we expect them to be able to use groupings of similar words to help with spellings, or to produce logical accounts of events. And so on.

It is not only in school that the children have to handle data. Playing games, making decisions on the predicted weather for the weekend, and sorting their collections of stamps, coins or soft toys all involve an understanding of some important aspects of data handling. In becoming competent and confident in handling data there are a range of experiences that the children need to have, starting with sorting out what is relevant according to appropriate criteria.

Early years work on sorting and classifying is the foundation for our handling data curriculum. In some cases such work may have to continue into later years for some children, for without the capacity to make sensible and reasonable choices much of what is expected in school will not be achievable. Moving on from this starting point the children need to learn how to organise data using appropriate tools such as tables and lists. To do this often involves them in counting and the organisation of counts, so devices such as tallying and the use of frequency tables and charts is essential too. Allied to these is the need to understand the differences between discrete and continuous data and that it is possible, within clear rules, to group frequencies and to develop cumulative totals. From this work there comes the need to present the outcomes of one's data collection and analysis.

The role of pictorial representation is clearly crucial and the children need to be able to select, from a range of possible ways of presenting their ideas, appropriate representations. Part of the way to achieve this is to analyse and interpret graphs, charts and tables that others present. Newspapers and magazines regularly provide their readers with information pictorially. To be able to detect bias in such presentations will help the children to avoid errors in their own presentations.

Finally, there is the whole area of probabilities that we need to introduce children to. Even where something is well-presented we still have to ask questions about its reasonableness. Children do need to be able to make reasoned judgements about the likelihood of events and the 'truth' of what they are told.

Simple charts

Learning targets

On completion of this theme the children should be able to:

1 ➡➡ tally
2 ➡➡ construct and interpret a block graph
3 ➡➡ interpret some bar charts.

Before you start

Subject knowledge

The ability to read and interpret charts and diagrams is of fundamental importance in a modern society. So much statistical information that has a real bearing on people's lives is represented pictorially that without this ability a person is seriously disadvantaged. Early work in handling data should have included simple mapping and the use of block graphs. Here we build on that early work. Tallying has a long history. It was, undoubtedly, the key way in which early humans kept records of possessions and engaged in trade. In fact the use of tally sticks in accounting was only phased out in the UK about 150 years ago. Tallying helps not only with the organisation of counts but also with later ideas about grouped data. Block graphs are a very common form of pictorial representation and one which children can readily continue to use as they progress through the middle parts of their primary education. Bar charts are like block graphs without the individual blocks. To read a bar chart it is necessary to consider height as the sole indicator whereas with block graphs it is still possible to extract information by counting on.

Previous knowledge required

Mapping, simple block graphs, 5× table, counting on particularly in steps other than 1.

Resources needed for Lesson 1

Access to a situation in which the children can do a count.

Resources needed for Lesson 2

Tube of Smarties®, Skittles® or similar – at least one tube for every pair of children, plain paper or paper towels, Copymaster 60.

Resources needed for Lesson 3

Copymaster 48.

Teaching the lessons

Lesson 1   ①

Key questions

Why use 5s?
How many are there altogether?

Vocabulary

Counting on, tally, total.

Introduction   [10 min]

▦ With the class, do a count of how many children there are. As you do this mark on the board the traditional form for tallying in 5s (卌). Talk with the children about the fact that counting in 5s and the use of this sort of tally mark has a long history and that tallying was used by early people as a device for keeping records of such things as flocks of sheep. Tally sticks were also a way of keeping records of agreed trades and so on. Show the symbol again, counting 1, 2, 3 and 4 for the vertical strokes and 5 for the diagonal stroke. Ask the children why this might be a good way of working. Get them to tally while you do another count of, e.g. the windowpanes or a set of books. Tell the children that they are going to work in groups to carry out a count using tallying.

Activities   [20 min]

▞ Depending on access and any available help choose a topic, or set of topics, for tallying. Examples of such topics include: tallying the children in other classes, the number of windows in the school, or the traffic passing the school. Each member of each group undertakes an individual tally and then the group has to agree a total.

Closing the lesson   [10 min]

▦ Give each group the opportunity to give the total they agreed and write them on the board in the form of a table. Ask what they found difficult, e.g. did they miss anything while they were recording? Keep the data collected for future use.

Assessment

Child's performance	Teacher action
Has problems with the process of tallying	Give more practice and check knowledge of counting on in 5s
Tallying not a problem	Move on to next lesson

Tallying not a problem   Move on to next lesson
and is clearly confident
in all aspects of this
work

Lesson 2 ②

Key questions

How many …?
What is the total?

Vocabulary

Block graph, total.

Introduction   10 min

Before the lesson get all the children to wash their
hands. This lesson should be a consolidation of earlier
experiences in the construction of block graphs.
Depending on the year in which you do this make
columns on the board with children's ages at the top
of each – have separate columns for boys and girls.
Ask for 'hands up' for each age and do a count – use
tallying as a reminder. Use the information to
construct simple block graphs for boys and girls. Ask
the children what characterises a block graph.

Activities   20 min

Move the children into pairs and give out paper towels
or plain A4 paper, tubes of sweets, and squared paper
(**Copymaster 60**). The children have to sort the
sweets they have in terms of colour and record these
in a table. Then, using the squared paper, they need to
construct a block graph of sweets in their colour sets.

Closing the lesson   10 min

Using some of the children's efforts talk through what
has been done and what might be concluded. If there
is time (or do this in a subsequent lesson) make a
block graph which totals the results for the whole
class. Save the information for the next lesson.
Depending on your judgement it might be
appropriate to let the children eat the sweets.

Assessment

Child's performance	Teacher action
Works steadily but does not complete task	Give more time for completion but check that the understanding is really there
Completes table and has clear idea about how to construct block graph	Move on to next lesson but give more time to complete block graph
Completes table and block graph	Move on to next lesson

Lesson 3 ③

Key questions

How does a bar chart differ from a block graph?
How many …?

Vocabulary

Block graph, total, bar chart.

Introduction   15 min

Using the work from the previous lesson remind the
children of the characteristics of a block graph. Now
explain that we do not always have to put in the
blocks, we can simply use a bar. Using the graph
below, work through the production of a *bar chart*,
explaining the term, on the board.

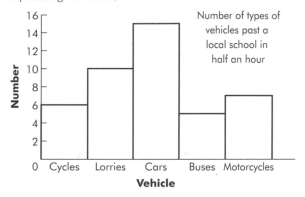

Number of types of vehicles past a local school in half an hour

Ask a variety of questions such as: What type of
vehicle passed the most/least? What 2 types total to
the same as cars? How many cycles and motorcycles
altogether? Point out that we can read off the totals
for each bar using the height of the bar – we do not
need to have blocks. Give out **Copymaster 48**.

Activities   20 min

The children need to work through the copymaster.
They can share ideas with each other and you,
especially when it comes to developing their own
questions.

Closing the lesson   10 min

Use the experience of the pairs here. What did they
find? Elicit some of the questions that the pairs
thought of and get the others to provide answers.
Collect in all the ideas with a view to collating the
questions for a follow-up session.

Assessment

Child's performance	Teacher action
Copes with some close support	Find more examples of bar charts and activities that will generate data that can be pictorially represented this way and give more practice
Completes questions satisfactorily	Give follow-up time to extend question set
Completes questions confidently and generates a number of own	Consider using child to collate the question bank

Homework

There are many opportunities here for data
collection. The children could tally such things as
mode of transport to school or their own journeys in
a week. They could tally numbers of potatoes,
carrots, baked bean tins and so on used in a week at
home. Get the children to recount any surveys on TV
or radio that they see or hear. For example, most
children ringing in thought that … was best.

THEME 35

Frequency and grouped data

Learning targets

On completion of this theme the children should be able to:

1 ➤➤ work with frequency charts and tables
2 ➤➤ undertake the grouping of some discrete data
3 ➤➤ interpret given pictograms.

Before you start

Subject knowledge

The idea of frequency is clearly linked to the collection and use of discrete data. Frequency is commonly depicted through the use of block graphs or bar charts in schools. Typical frequency work involves the collecting of data about observable or easily quantified responses, e.g. numbers of people in a family, favourite colours or TV programmes, and numbers of children born in particular months or seasons. These examples are all of discrete data – that is each is clearly separate from the other members of a set. If we were surveying pets, say, then one could own 2 hamsters and then acquire a third but 2.5 pets is not possible. Each is a separate entity; and hamsters are not cats, though both are animals.

Sometimes the numbers of items are such that it makes sense to allocate them within a chosen group. For example, a survey of numbers of felt-tip pens owned by children would probably range from some with 2 or 3 to those with large collections of 48 or more. In these circumstances

it would make sense to group the results of a survey in terms of, say, 0–9, 10–19, 20–39 and so on. Note that the groups must not overlap. It is a common mistake to group like this: 0–5, 5–10, 10–15. In grouping our felt-tips data we are, of course, using 'grouped discrete data'. Such grouping underpins the pictorial representation of the pictogram.

Previous knowledge required

Block graphs, collection and use of discrete data, tallying.

Resources needed for Lesson 1

Copy of Graph A (below) on an OHP transparency or the board, Copymaster 60.

Resources needed for Lesson 2

Copymaster 60.

Resources needed for Lesson 3

Copymaster 49, an example of a pictogram from a newspaper or another mathematics resource ideally transferred onto an OHP transparency or copied onto the board before the lesson.

Teaching the lessons

Lesson 1 ①

Key questions

How often?

Vocabulary

Frequency, table, quantity, total, amount.

Introduction [15min]

▓ Either on the board or an OHP show the children the frequencies depicted in Graph A. Talk about 'how often?' or 'how frequent?' are the quantities. Tell the children that they are going to acquire some information about frequency, working in small groups.

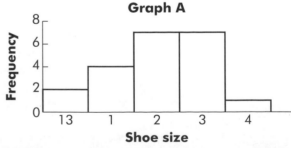

Graph A

Frequency (vertical axis): 0, 2, 4, 6, 8
Shoe size (horizontal axis): 13, 1, 2, 3, 4

Activities [25min]

⚀ The children have to collect frequency data on one or more topics depending on whether you want the class to investigate one idea or each group to attack a different idea. Ideas for topics are:

a Number of times each of the vowels, or the letters of someone's name, appears on a book's page.

b Number of people in children's families.

c The 4 seasons and when your birthday falls (divide the year as March–May, spring; June–August, summer; September–November, autumn; December–February, winter).

d How many times a week do you eat/drink …?

e How frequently does the school make use of TV/radio in a week?

f Shoe sizes in the class.

g Frequency of eye colour/left-handedness.

Closing the lesson   `10min`

Select 1 or 2 examples of what the children have been doing and share these with the class. Give additional follow-up time for completion of tables, etc. and then collect in all the information generated for use at a later stage.

Assessment

Child's performance	Teacher action
Collects some data but clearly needs more data collection time	Give more time and opportunity but be prepared to revise work on block graphs
Collects the data and clearly knows how to produce a frequency table	Move on to next lesson
Collects data and makes good progress in organising what has been found	Move on to next lesson and get child to help others at a later stage

Lesson 2   ②

Key questions

How can we find this out?
How shall we group our data?

Vocabulary

Data, groups, grouped data.

Introduction   `15min`

Tell the children that they are going to collect some information (data) about how long it takes each person to get to school in the morning. It does not matter how they travel. We are just interested, on this occasion, on how long it takes in minutes. Tell them that their work, if well done, will be of considerable interest to the Head and the Governors. Firstly ask them to decide how to collect the data needed. Get the children to make suggestions and come to an agreement with them on how to proceed.

Activities   `25min`

Using the agreed approach data is collected. Get the class together again and put the discrete data on the board. Tell them that with so much discrete data it is useful to group times together, e.g. less than 5 minutes, 5–9 minutes, 10–14 minutes and so on. The data grouping will depend on your particular circumstances. Working in small groups the children must now put the data into the agreed groupings and then draw a block graph or bar chart using squared paper.

Closing the lesson   `10min`

Sketch an example graph or chart on the board and ask whether they have produced something like this. Decide whether you could proceed to a whole school survey at this stage.

Assessment

Child's performance	Teacher action
Participates fully but finds some of the task confusing	Give more practice
Participates fully and makes good progress	Move on to next lesson
Completes all of the main elements of the task successfully	Move on to next lesson but involve all the children in the whole school survey if you go on to do this

Lesson 3   ③

Key questions

What does each picture stand for?
How many?

Vocabulary

Group, pictogram.

Introduction   `10min`

Use the pictogram you have found to discuss what it is about, what the symbols mean, and what we might learn from the presentation. Tell the children that this sort of representation is known as a *pictogram*. Tell them that they are going to look at 2 more examples, working in pairs, writing out all the things they can find out from the given pictograms. Give out **Copymaster 49**.

Activities   `20min`

The children work on the copymaster, sharing all their ideas and listing those they are happy with.

Closing the lesson   `10min`

Get the children to tell you, taking an idea from each pair in turn, what they have found out. Make a note of the ideas on the board then recap on them.

Assessment

Child's performance	Teacher action
Comes up with some ideas	Give more pictograms to explore
Comes up with several ideas	The learning targets for this theme have been met
Has many ideas	The learning targets for this theme have been met

Homework

Ask the children to make a collection of pictograms, block graphs and bar charts they come across with the help of their family. Use the surveys done to produce tables, charts or graphs. Give some of the ideas not used in the lessons as individual projects.

THEME 36 Graphs

Learning targets

On completion of this theme the children should be able to:

1 ➤➤ construct and interpret a bar line graph
2 ➤➤ construct and interpret a line graph
3 ➤➤ interpret simple pie charts.

Before you start

Subject knowledge

The work in this theme is intended to add to the repertoire of the children in presenting and interpreting the pictorial representation of data. The bar-line graph builds on their knowledge of block graphs and bar charts. It also acts as a good intermediate step to graphs constructed on the basis of co-ordinates. Line graphs are important tools not only in handling such things as survey data but also in understanding relationships. The line graph that shows fluctuations is only appropriate when the x-axis is based on measures of time. Examples of this would be temperature over a year in a location, or fluctuations in income and outgoings over a period of time. Line graphs should be used for continuous data, not discrete data. Pie charts, though highly visual, are difficult to construct by hand, needing a good understanding of angle and proportion, and skilled measurement. Many software packages will produce pie charts for you so this is now much more accessible to primary school children.

Previous knowledge required

Block graphs, bar charts, tallying frequency, fractions, temperature.

Resources needed for Lesson 1

Examples of block and bar graphs, travel/holiday brochures which give information about temperatures at different locations and times of the year, atlases, rulers, Copymaster 71.

Resources needed for Lesson 2

The work done in Lesson 1 as well as the same resources.

Resources needed for Lesson 3

Copymaster 50.

Teaching the lessons

Lesson 1 ①

Key questions

Where must we write the months of the year?
How tall should this line be?

Vocabulary

Block graph, bar chart, bar-line graph, temperature.

Introduction 15 min

▓ Using examples of block graphs, either child-produced or from other texts, remind the children of what such graphs do and how they are constructed. Move on to an example of a bar graph and explore its characteristics. Explain that it is possible to make a graph using lines rather than blocks or a bar. Turn one of the examples you have used into a bar-line graph on the board. Show how to establish the length (height) of each line and the fact that the line is labelled not the space between lines. Tell the children that they are going to produce a bar-line graph on temperature with months along the x-axis and temperature on the y-axis, and draw the axes on the board.

Activities 25 min

▐▐ Working collaboratively using the brochures you have assembled, squared paper (**Copymaster 71**), pencils and rulers, and atlases (so the children can see where in the world they are looking at temperatures), the children have to produce a labelled and named bar-line graph.

Closing the lesson 10 min

▓ Discuss with the children what they feel they have learned. Explore any connections they make between block or bar graphs and bar-line graphs. Collect in the work and save it for the next lesson. If anyone has not finished then give them time before the next lesson. Keep all the brochures.

Assessment

Child's performance	Teacher action
Makes slow progress	Check to see whether the issue is speed or understanding; give more time with close support
Produces a bar-line graph after a few attempts	Move on to next lesson but revise the work beforehand
Produces a bar-line graph readily	Move on to next lesson

Lesson 2 ②

Key questions

What do the points represent?
What does the line represent?

Vocabulary

Bar-line graph, line graph.

Introduction ⌊10 min⌋

Remind the children of the work they did on the bar-line graph of temperature in the last lesson. Tell them that in this lesson they are going to use the same data to produce another sort of pictorial representation – a line graph. Use the example in Graph B to show the sort of line graph you are looking for. Make connections with bar-line graphs: here we just have the points which have to be accurately placed.

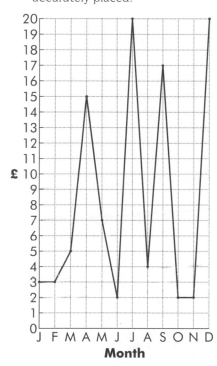

Graph B
A line graph of my income over a year. My birthday is in July and relatives visited in April and September. Of course an 'outgoings' graph is also needed.

Activities ⌊25 min⌋

Using the holiday brochures as before, the children now have to produce a 'spiky' line graph of temperatures over the year. Keep reminding them of the need for accuracy in placing each point before they join them up. If necessary give those children who need extra help the bar-line graphs they produced from the last lesson.

Closing the lesson ⌊10 min⌋

Talk over the ideas behind line graphs of this sort. Stress the need to have the x-axis based on time and the need for continuous and not discrete data.

Assessment

Child's performance	Teacher action
Completes the graph but needs a lot of support and encouragement	Further practice may be needed; consider revising all aspects of pictorial representation to date
Completes the graph with some support	Move on to next lesson
Completes the graph with little support	Move on to next lesson and consider getting child to help with a display of the class efforts

Lesson 3 ③

Key questions

What do these mean?
What is the most common/popular?
What is the least common/popular?

Vocabulary

Common fractions, pie chart.

Introduction ⌊10 min⌋

Draw the first of the pie charts in the diagram below on the board. Take the children through the pie chart, giving its name and asking why it is so called. Ask 1 or 2 questions to help orientate everyone then draw the second pie chart and ask the children to explain what it is about. Give out **Copymaster 50**.

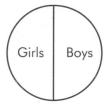
Children in school

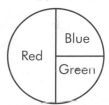
Colour of exercise book that a company produces

Activities ⌊20 min⌋

The children work through Copymaster 50 producing as many statements as they can. These should be recorded on the copymaster.

Closing the lesson ⌊10 min⌋

Sample from the children's ideas and make notes of these on the board. If there are any novel ones get the children to explore them further. Recap what the advantages and disadvantages of pie charts are.

Assessment

Child's performance	Teacher action
Can make simple but accurate statements	Give further opportunities to work with simple pie charts
Develops a range of ideas using appropriate vocabulary	Consider giving opportunities to construct own simple pie charts
Clearly understands some of the main underlying principles, such as proportion	Give opportunities to construct own simple pie charts

Homework

Collect examples of line graphs and pie charts from the newspapers with family help. Look for sunrise and sunset data in the papers and plot these on a line graph over, say, a period of 4 weeks.

THEME 37 Averages

Learning targets

On completion of this theme the children should be able to:

1 ➨ Explain what 'average' is and that there are different ways of arriving at an average
2 ➨ Calculate the mean average for given examples
3 ➨ Talk about the idea of 'range'.

Before you start

Subject knowledge

There are 3 averages that the children need to know about. These are the *mean*, the *mode* and the *median*. The mean average is the one that most people think about and use in everyday discussion. This is the average where we total all the quantities and divide by the number of quantities. However, the mean average is the most difficult to determine and is not always the most helpful average to use. For example, if a shoe manufacturer was to measure many people's feet and then take the mean average it is quite likely that only a few of us would actually find shoes to fit us. What the shoe manufacturer needs to do is to look at the most common sizes and manufacture numbers of shoes that correspond to this. This is using the modal average – the most common. If we looked at, say, a set of annual salaries of a few people we can compare the difference between them using mean and mode. The salaries shown below are deliberately chosen to show up the differences though such disparities can and do occur. Clearly the modal average is a more useful judgement than using the mean average.

Five people in our road have these annual salaries:
£90,000, £26,000, £12,000, £12,000 and £10,000

The total of these is £150,000, and the mean average is:

$$\frac{150,000}{5} = £30,000$$

but only 1 person earns more than this.
The modal average is £12,000 which looks to be a more useful judgement.

The median average is where we take the middle quantity. It is easy to determine, compared with the mean, and often gives us a good feel for the results of an event. To work out the median put the quantities in order of size then take the middle one as the average. Where there are an even number of results then we take the half-way position between the 2 central quantities. The median average is often used in tests when we are looking at the top and bottom halves of a group, and the top and bottom 25% of cohorts.

Maths test marks for 9 children:
20, 18, 3, 1, 9, 7, 14, 15, 12
Reorder them:
1 3 7 9 12 14 15 18 20
↑
The median average
(The mean average is 11)

English test marks for 6 children:
8, 8, 11, 5, 7
Reorder them:
1 1 5 7 8 8
↑
The median is half way between 5 and 7 (6)
(The mean average is 5)

Previous knowledge required

Division, calculator use, counting and ordering

Resources needed for Lesson 1

Copymaster 51.

Resources needed for Lesson 2

Copymaster 52, calculators.

Resources needed for Lesson 3

Copymaster 53.

Teaching the lessons

Lesson 1 ➊

Key questions

What do you think 'average' means?
Why are 'averages' useful?

Vocabulary

Average, mean, mode, median.

Introduction 20min

▓ Start by asking the children if they have heard of 'average' and ask what they think it might mean. Typical examples might include sports averages, average in a school context, and average as a term for not good/not bad. Explain that there are 3 kinds of average and that they are called mean, mode and median. Tell the children that they will work on mean average next time but for now we are concentrating on the other 2. Using the examples given above, or others you have available, explain how we obtain the median and modal averages. Give the children ideas about where these averages might be useful. Give out **Copymaster 51**.

Activities | 20min |

The children attempt the challenges on Copymaster 51. Go round the class reminding them of the ways in which these averages are determined.

Closing the lesson | 10min |

Recap on the processes involved and, if time, agree some of the answers to the copymaster challenges.

Assessment

Child's performance	Teacher action
Finds averages difficult	Work through the ideas again and then retry the copymaster
Needs more time on the copymaster	Give time to complete lesson/copymaster
Completes all main parts of the task satisfactorily	Move on to the next lesson

Lesson 2 ②

Key questions

How can we work this one out?
What do we have to do?
What does this tell us?

Vocabulary

Average, mean, mode, median.

Introduction | 15min |

Remind the children of what they did last time and get them to tell you what median and modal averages are. Explain that mean average is one that most people think about when we use the term 'average'. Use the example below and work it through on the board.

> Numbers for showing how to work out the mean:
> 6, 9, 13, 5, 5, 5, 6
> If you feel it is appropriate point out the modal average.

Activities | 20min |

Using the challenges on **Copymaster 52** the children have to work out the mean averages in each case. Where a calculator can be used this is marked on the copymaster.

Closing the lesson | 10min |

Recap the process of calculating a mean average and finish by explaining that mean average is not the whole story. We need to know some other things including the range of quantities.

Assessment

Child's performance	Teacher action
Completes some of the work	Give more time but go over the processes again

| Completes most of the work | Give more time and check the final responses |
| Completes all of the work | Move on to next lesson |

Lesson 3 ③

Key questions

What averages can you work out?
What is the range here?

Vocabulary

Range, average, mean, mode, median.

Introduction | 10min |

Working out the range is straightforward; the problem is in understanding its significance. Start by putting up the first set of marks in the box below and show how we determine the range. Then put up the second set on the board and get the children to tell you the range. Now point out that the mean average for each set is the same but the ranges are very different.

> First set of marks:
> 2, 5, 7, 9, 11
> The range is either
> $11 - 2 = 9$ or 2 to 11 which is 10.
> Either can be used.
> Second set of marks
> 1, 4, 8, 14, 7

To judge whether we are comparing similar things when we look at a mean average we really also need to know the range. Give out **Copymaster 53**.

Activities | 20min |

The children work through the challenges on the copymaster.

Closing the lesson | 10min |

Recap the ideas the children have experienced in this theme on all the 'averages' and the additional information that range gives us.

Assessment

Child's performance	Teacher action
Completes some of copymaster	Give support and more time; check on understanding and process of determining 'average'
Completes most of copymaster	Give the chance to finish and then check with the child present
Completes all the copymaster challenges	The learning targets for this theme have been met

Homework

Get the class to make contributions to an 'average' display board. Wherever they meet someone using the term 'average' they should make a note of it or cut it out from the newspaper or magazine and bring it to school for the display.

THEME 38 | Chance

Learning targets

On completion of this theme the children should be able to:

1 ➤➤ make decisions about how important 'chance' is in given situations
2 ➤➤ work with simple probabilities
3 ➤➤ talk about the idea of 'certainty'.

Before you start

Subject knowledge

The idea of 'chance' which is often expressed in everyday discussion as 'fate' or 'luck' is a core idea for people. It is the case that millions play the National Lottery hoping for some luck even though the odds are so much against them winning. In working with children there are a number of components of any discussion about chance and probability that we need to provide. Key ideas are randomness, certainty, sampling, and prediction. In this theme, 3 related topics, which are important facets of these ideas, are developed: chance, probability, and confidence.

Here we are looking at the element of chance within common games as these can allow us to focus clearly on all the important aspects of 'luck'. Through simple probability work we can see that chance can be assessed and that by employing such assessments we can form judgements about the likelihood of events. It is important to distinguish, though, between theoretical odds and practical outcomes. For example, in Lesson 2 the children are rolling conventional dice. In theory the chance of any of the numbers on a die coming up is 1 in 6 (the odds against a 6 is therefore 5:1 against). In practical tests it is often the case that equal totals of each of the numbers are not achieved, though you should get totals close to those that can be theoretically predicted. This theme offers some good opportunities for creating displays of, for example, games evaluations.

Previous knowledge required

Playing of games which use dice, simple fractions or proportions, the use of tables.

Resources needed for Lesson 1

As many examples of board games as you can muster (invite the children to bring in some), books on games – ancient and modern.

Resources needed for Lesson 2

Traditional dice enough for 1 between 2.

Resources needed for Lesson 3

Copymaster 54.

Teaching the lessons

Lesson 1 ①

Key questions

How do we play this?
What are the important rules?
What does 'lucky' mean?

Vocabulary

Chance, luck, strategy.

Introduction 10min

Ask the children to give you some examples of when people talk about chance. What do they think people mean by such terms as 'luck' or phrases such as 'they haven't got a chance'? Tell the children that in this lesson they are going to look at how a range of games is played. The challenge is for them to decide how much of the winning of a game is dependent on chance and how much on strategy. Organise the distribution of games to the groups as appropriate.

Activities 25min

The children explain, explore and try out games but think about questions to do with chance as they do so. They should keep notes on what they discover.

Closing the lesson 15min

Get each group in turn to give some feedback in respect of how they rate the games on chance against strategy. See if you can develop some ranking from those that depend entirely on 'luck', like snakes and ladders, and those that are strategic, like chess.

Assessment

Child's performance	Teacher action
Enjoys the games but finds it hard to make decisions about 'chance'	Give some more practice concentrating, in the first instance, on dice games
Is able to state some facts about the level of 'chance'	Move on to next lesson
Can organise ideas in making comparisons	Move on to next lesson

Lesson 2 ②

Key questions

What are the chances?
What is the total?
How do your predictions match the overall outcomes?

Vocabulary

Chance, table, total, comparison, prediction, outcome.

Introduction 〔10min〕

 The children working in pairs are going to roll dice and record the numbers that come from each throw. Explain this to them but first they have to predict what the result might be. Discuss this with them and then move them on to the activity. The children should throw and record at least 12, but preferably more, throws.

Activities 〔20min〕

 The children have to make a table and roll their dice at least 12 times, keeping a record of the result of each throw.

Closing the lesson 〔15min〕

 Make a table on the board and get each pair to give you their results. Total the results and see what the distribution of 1s, 2s and so on is. Discuss the results in the light of the prediction made at the start of the lesson. How close are they?

Assessment

Child's performance	Teacher action
Manages to record accurately at least 12 throws	Reassure yourself that the process was understood; check through understanding of the prediction against the class results
Manages to organise and keep records of more than 12 throws	Move on to next lesson taking the opportunity to ask about understanding of the prediction against the class results
Manages all of the task and makes a strong contribution to the discussion	Move on to next lesson

Lesson 3 ③

Key questions

How likely is this?
What is the likelihood that …?

Vocabulary

Chance, likely, likelihood, event.

Introduction 〔15min〕

In this lesson we are concerned with making a step towards later use and understanding of the probability line. The quantification of probabilities will come at a later stage; here we are looking to develop a likelihood line. Draw a line like the one on **Copymaster 54** on the board. Get the children to make suggestions as to some things they can predict and with their help locate them on the line. If you need to suggest ideas use the weather, holidays, and meeting famous people to generate discussion.

Activities 〔20min〕

Give out the copymaster. The children have to discuss and agree events and their location on the likelihood line. Individuals should keep a record on their copy of the copymaster.

Closing the lesson 〔15min〕

Elicit some of the ideas those different pairs or groups have had and put them on the line on the board. There may well be disagreements about likelihood and these should be pursued. Collect in individual responses and consider making a class likelihood line as part of a display on probability and chance.

Assessment

Child's performance	Teacher action
Puts some ideas down but mostly these are derived from the introduction	Go over the work again and let child have another go; consider getting those who found the idea fairly straightforward to help in a peer review setting
Develops some ideas in addition to those on the board	The learning targets for this theme have been met
Develops a range of ideas many of which have a degree of originality	Consider child helping to set up and co-ordinate a display using the class ideas

Homework

The children could try out games with their families and get opinions from friends and relatives on the amount of strategy there is in games. In some cases children might be able to evaluate computer games too. Coins were not used in the lesson as dice are less likely to cause disruption but the children could try tossing a coin at home and bringing their record of throws to school to pool with results the other children have provided.

Co-ordinates

Learning targets

On completion of this theme the children should be able to:

1 ➨→ produce line graphs from given information
2 ➨→ use the idea of continuous data to establish intermediate points on a line
3 ➨→ produce linear conversion graphs.

Before you start

Subject knowledge

It was Descartes who first combined algebra and geometry and it is after him that Cartesian co-ordinates are named. The key feature of Cartesian co-ordinates is the ability to express precisely a point. Take, for example, map references, using Ordnance Survey maps where it is possible, because of the co-ordinate structure, to locate very precisely a point on the map. Using algebra it is possible to produce equations that describe relationships which can be graphed using co-ordinate graphs. For example, in $y = x + 1$ the numbers from, say, 1 and 10 can be substituted for x giving values of y of 2 and 11 respectively. The points described by 1, 2 and 10, 11 can be marked using the x-axis and y-axis of a co-ordinate graph and a straight-line graph is produced. All the points on this straight line will conform to our equation. This means that we can determine not only values for y (or x) in respect of whole numbers but we can also determine intermediate values.

Previous knowledge required

Pictorial representation, use of grid squares like those on some town maps which use labels such as B3 to identify a square in which a street or building is located, multiplication tables, drawing skills, function machines.

Resources needed for Lesson 1

Graph paper as on Copymaster 71, rulers, sharp pencils.

Resources needed for Lesson 2

Graph paper as on Copymaster 71, rulers, sharp pencils.

Resources needed for Lesson 3

Graph paper as on Copymaster 71, rulers, sharp pencils, Copymaster 55.

Teaching the lessons

Lesson 1 ①

Key questions

What is the function?
How are these related?

Vocabulary

Function, graph, co-ordinates, relationship, position.

Introduction ⬚15min

 The class may have done some graphing of multiplication tables previously. Remind them of this and with their help draw one on the board like that shown opposite for the 3× table. Tell the children that you want them to produce multiplication graphs for the 2× and 4× tables on the same piece of graph paper.

Activities ⬚25min

👤 Go around the class helping as the children work on their graphs. Encourage the children to talk about any ways they can see of making the drawing of such graphs simple.

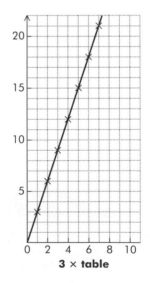

3 × table

Closing the lesson ⬚10min

 Draw the graphs the children should have obtained on the board so that they can self-check. Emphasise that the graphs are straight-line graphs, and that in such cases you only need 2 points to be able to draw the graph and then use it. This will be picked up again when looking at intermediate points in the next lesson.

Finish by pointing out that we can make general statements for each of the lines on the graph: for the 2× we could state that $y = 2x$ and for the 4× that $y = 4x$.

Assessment

Child's performance	Teacher action
Part of the activity is finished	Check on understanding of how to use co-ordinates to identify points – child may still be thinking of grid square identification; then repeat exercise as appropriate
Most of the activity is completed	Allow time to finish then check before moving on to next lesson
All done successfully	Move on to next lesson

Lesson 2 ②

Key questions

What would … cost?
What is y when x is …?

Vocabulary

Function, graph, co-ordinates, relationship, position, intermediate value/point.

Introduction 15 min

 Quickly go over the main points of the previous lesson with the help of the children. Tell them that you are planning a party for all your many relatives and they like crisps and lemonade. Your home-made lemonade costs 15p a bottle to make. Draw a graph on the board to show how you can work out the cost of different numbers of bottles. Tell the class that they are going to help you by producing a graph about crisps which you cannot make and will have to buy. One family pack of *Roast Ox* which is the relatives' favourite costs £2.20. There are 10 packets in each family pack. Ask the children to draw a graph on the basis of this information – you will ask some questions later. Talk about the need to be thoughtful about which axis to use for packs and which for money.

Activities 25 min

 Give time for the pairs to draw up their crisps graph. When it is appropriate stop them and put some questions on the board. These should include items like: What will it cost me if they eat 7 packs? What if they eat 45 packets? How many packs or packets did they eat if it costs me £7.04? Let the children come up with answers in their pairs.

Closing the lesson 10 min

Using the graphs and the questions go over the answers with the children's help. Ask them how they worked out the cost when it was on the line but not at a point they had drawn in. Reinforce the fact that we can read off a line graph either x or y values for any point on the line.

Assessment

Child's performance	Teacher action
Completes the graph but struggles with your questions	Spend some time in discussion, asking further questions and helping to elicit answers
Completes both tasks but makes errors in response to some of the questions	Check the kinds of errors made with child and use this to decide whether more work on, e.g. manipulation of decimals, is needed
Completes both tasks successfully	Move on to next lesson

Lesson 3 ③

Key questions

What is this in °F?
What is this in °C?

Vocabulary

Function, graph, co-ordinates, relationship, position, intermediate value/point, temperature language.

Introduction 10 min

 Quickly recap what the children should already have learned about straight-line graphs. Tell them that they are going to produce a conversion graph in this lesson. Show them a miles-kilometres conversion graph on the board. Ask if anyone has noticed that in weather forecasts the forecasters give temperature in °C and °F. Explain that Fahrenheit is an old UK measure that many people still use. Give out **Copymaster 55**.

Activities 25 min

 The children work through the copymaster plotting the graph and answering the questions.

Closing the lesson 10 min

Use individual responses to discuss findings and to gain an impression of how the children have done.

Assessment

Child's performance	Teacher action
Produces graph with help	May need to go back over some of the earlier work
Produces graph but makes some errors in answering questions	Check child is clear about the way to read one axis against another
Completes all of task satisfactorily	The learning targets for this theme have been met

Homework

Make a single graph on which all tables to 10×10 appear. Make straight-line graphs of $y = x$, $y = x + 1$, $y = x + 2$ and so on. What do they notice? What about $y = x - 1$? Produce their own copy of the miles–kilometres conversion graph. Try to produce a speed conversion graph, or pounds and kilograms, or inches and centimetres.

Fair or unfair?

Learning targets

On completion of this theme the children should be able to:

1 ➤➤ make statements about fairness in relation to particular games
2 ➤➤ use previous learning to make decisions about fairness
3 ➤➤ detect errors in given pictorial representation examples.

Before you start

Subject knowledge

All of us, and children particularly, have a well-developed sense of fairness. Usually we express this in the context of justice and injustice. Here we are concerned with a sub-set of the notion of justice – a sub-set that we can get a view on when thinking mathematically. In discussing probabilities and whether the odds are even or favour one type of outcome, we are using powerful mathematical ideas. Whilst it may not be possible or even desirable to suggest that all decision making can be statistically analysed we can suggest to children that in certain circumstances it makes a lot of sense to evaluate the probabilities of different outcomes. In this theme we take the opportunity to make reasoned judgements about some outcomes from games, and get the children to see that even when data is presented pictorially we have to decide whether it is a reasonable representation.

Previous knowledge required

Practical work in mathematics including experience of investigations, familiarity with the main types of pictorial representation.

Resources needed for Lesson 1

A pair of dice for each child.

Resources needed for Lesson 2

Dice for each pair of children, Copymaster 56.

Resources needed for Lesson 3

Copymaster 57.

Teaching the lessons

Lesson 1 ①

Key questions

Is this fair?
What are your reasons for stating that this is fair/unfair?

Vocabulary

Fair, fairness, unfair, chance.

Introduction 10 min

Tell the children that they are going to play a game and investigate some aspects of how it works. First they have to try the game, then decide whether it is fair to all players or not, and be able to explain their decision. If they think it unfair they have to suggest ways in which the game might be adapted to make it fairer. Give each child a set of dice and divide the class into pairs telling each pair which of them is Player 1 and which Player 2. Explain the rules of the game. They each take turns to roll their two dice. If the total is 2, 3, 4, 10, 11 or 12 then Player 1 gets a point. If the total is 5, 6, 7, 8 or 9 then Player 2 gets a point. The first to 25 points wins. Point out that Player 1 has 6 winning totals and Player 2 only 5.

Activities 25 min

The children play the game. They should find that Player 2 is usually the winner. (With any game it is always possible for any player to win even when the odds are very much against them.) Ask why Player 2 seems to be winning even though they have fewer totals with which to score a point? How could the game be made fairer? The children should formulate their responses as a pair.

Closing the lesson 15 min

Sample the responses from a range of pairs. There are a variety of ways in which they might present their answers. If it proves difficult, or as a way of summarising, put the table below on the board and talk it through with the help of the children. You might choose to pursue the patterns in the table on another occasion.

Dice (+)	1	2	3	4	5	6
1	2	3	4	5	6	7
2	3	4	5	6	7	8
3	4	5	6	7	8	9
4	5	6	7	8	9	10
5	6	7	8	9	10	11
6	7	8	9	10	11	12

Assessment

Child's performance	Teacher action
Plays game and feels it unfair but cannot analyse why this is so	Give more opportunities for games playing using dice, spinners and playing cards; talk through the outcomes of these games
Plays game and makes suggestions for adaptation but does not offer thought out analysis	Move on to next lesson but try to give another dice investigation first
Completes all of task giving a good explanation as to why Player 2 is favoured	Move on to next lesson

Completes all of task giving a good explanation as to why Player 2 is favoured and how the game might be made fairer	Move on to next lesson

Lesson 2

Key questions

Is this fair?
How could you make this fairer?

Vocabulary

Fair, fairness, unfair, chance.

Introduction `15min`

 Start with a discussion using a number of examples of situations where fairness and unfairness should be considered. Examples include how a raffle works and what would make it unfair, and why runners start at different places on the track in some races. Get the children to tell you a rhyme for choosing someone for a team. Can they make it work out the way they want? How? **Copymaster 56** is based on work done in The Netherlands on children's perceptions of fairness.

Activities `25min`

 Group the children in pairs telling them who is Player 1 and who is Player 2. Give out the copymaster. Player 1 has a dice and moves along the track according to the number they roll on their turn. Player 2 always moves 4 spaces. The children play the game several times. Player 2 usually wins. Ask them to work out an explanation as to why this is the case. Could the game be made fairer?

Closing the lesson `10min`

 Use ideas from selected pairs to explore the fact that Player 1 has only 2 out of 6 numbers that will exceed 4.

Assessment

Child's performance	Teacher action
Plays game and feels it unfair but cannot analyse why this is so	Give more opportunity to play and research similar games
Plays game and suggests an explanation as to the outcomes	Do some work on probabilities with dice then move on to next lesson

Lesson 3

Key questions

What is misleading about this?
What else do you need to know?

Vocabulary

Terms relating to all aspects of common pictorial representation, misleading, continuous data, discrete data.

Introduction `10min`

 Draw axes on the board and draw a bar chart with 5 bars of different height. Do not label the diagram. Ask the children what sort of a graph it is and then what else they would need to know to interpret it. Tell them it is a graph of favourite pop stars and write this title on the board. Take further suggestions as to what they need to know to fully understand the graph. Tell them that they are going to work in small groups and look at some more graphs that have things wrong with them.

Activities `25min`

 Give out **Copymaster 57**. The children have to decide what is wrong or misleading in each graph. Get them to share their ideas and to keep notes for the end of the lesson.

Closing the lesson `15min`

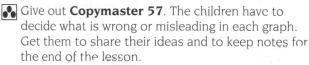 Using their notes the children should pick up on the errors. These include widths of bars, inappropriate use of a line graph for discrete data, lack of labels, and absence of overall sample sizes.

Assessment

Child's performance	Teacher action
Spots some errors	Give more practice on interpreting different sorts of representation
Spots most errors	Repeat the exercise at a later date
Spots all errors	The learning targets for this theme have been met

Homework

Invite the children to try out some of the games at home and collect families' and friends' reactions. Explore mathematics reference books, mathematics investigation books, and books on games to find other examples. Use exercises from mathematics texts available to you on simple probabilities, and/or the interpretation of pictorial representations.

THEME 41 | Probability

Learning targets

On completion of this theme the children should be able to:

1 ➤➤ work with the probability line
2 ➤➤ develop ideas about the range of outcomes from given events
3 ➤➤ construct, and explore parts of Pascal's Triangle.

Before you start

Subject knowledge

The use of a probability line is difficult and should not be attempted too early in a child's work on probability and chance. It helps to think of marking out of 100 (a percentage) when first judging probabilities. These percentages can then be divided by 100 to give values between 0 and 1 for the probability line. In making judgements it is important to discuss the subjective nature of these. The probability line should also be built on earlier experiences of a likelihood line. In cases where we can actually measure outcomes, both theoretically and practically, the use of quantities is not subjective – though prediction may be. For example, in tossing 2 coins we can work out that there are 4 possible outcomes and that the chance of 2 heads is 1 in 4. If we were expressing this as odds we would say that the odds are 3 to 1 against getting 2 heads (the 1 + 3 making our 4 possibilities). The outcomes for 2 and 3 coins are shown here.

If we total the heads, head/tail combinations, and tails for the 2 coins we get a 1:2:1 distribution. For the 3 coins we get a 1:3:3:1 distribution.

Possible outcomes with 2 coins:

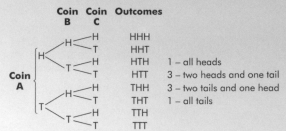

Possible outcomes with 3 coins:

These distributions will be picked up when tackling some aspects of Pascal's Triangle. Blaise Pascal was a 17th century French mathematician.

Previous knowledge required

Number lines, likelihood lines, number patterns work, early work on probability and chance.

Resources needed for Lesson 1

Copymaster 58, a probability line on a board.

Resources needed for Lesson 2

Coins – 3 of the same denomination per group.

Resources needed for Lesson 3

Copymaster 59, Multi-link® or similar – 6 cubes for each small group, 3 of one colour and 3 of another.

Teaching the lessons

Lesson 1   ①

Key questions

What are the chances?
What is the probability that …?

Vocabulary

Probability, chance, likelihood, probability line, comparison, judgement.

Introduction |15 min|

▓ Draw a probability line on the board and mark it from 0–1. Put a statement on the board that represents a

likely event – the weather is a good source. Give your statement a mark out of 100 and see if the children agree. When there is consensus tell them that if we now divide the mark by 100 then we have a decimal fraction which will fit on the line between 0 and 1. Locate it and add a marker for 0.5 stating that this is a 50% chance and if we divide 50 by 100 we get our 0.5. Give other examples if you feel it necessary.

Activities |25 min|

▤ Give out **Copymaster 58**. The children work individually through the given events, then create a few ideas of their own. As they complete this progressively form small groups so that they can share their ideas. Do they agree with each other?

Closing the lesson 10min

 Sample some of the given and created events and the judgements about their probability, and discuss these.

Assessment

Child's performance	Teacher action
Has some difficulties with turning 'marks' into decimals	Go over decimal fractions and percentages again
Works diligently but clearly finds it demanding	Give more opportunities
Has good ideas and listens to others	Engage child in development of the display

Lesson 2 ②

Key questions

How many of this combination are possible?
Can you work out all of the possible outcomes?

Vocabulary

Outcomes, chance, odds.

Introduction 15min

 Start by asking the children what the chances are of a head or a tail when you toss a coin – remind them of the probability line and where this event might appear. What are the chances of rolling a 6 with a conventional dice? What are the chances of turning up a spade in an ordinary pack of cards? An ace? The ace of spades? Tell them you are now interested in what might happen when you combine events.

Activities 20min

 Give each pair 2 coins of the same denomination. By trial and discussion you want to know what are the chances of tossing 1 coin then the next and getting 2 heads. When the children have attempted this put a copy of the diagram 'Possible outcomes with 2 coins' (opposite) on the board.

Some children may think there are only 3 possibilities. Spend time explaining that tail-head is a different combination from head-tail. Give out an extra coin to the pairs and ask them to try and work out all the possibilities for 3 coins.

Closing the lesson 10min

 Ask the children what they think are the possibilities of getting 3 heads, or 3 tails, when tossing 3 coins. This information will be useful in the next lesson.

Assessment

Child's performance	Teacher action
Manages the 2 coins but not the 3	Go over the ideas again then move on to next lesson
Manages the 2 coins and makes some progress with solving for 3	Move on to next lesson
Solves for both 2 and 3 coins	Move on to next lesson

Lesson 3 ③

Key questions

What patterns can you see?
How does this connect with your other work on probability?

Vocabulary

Pascal's Triangle, probability, chance, pattern, sequence.

Introduction 5min

 Remind the children of the work they did on coins. Ask them if they remember the pattern for 2 coins, and then 3 coins.

Activities 30min

Give the Multi-link® cubes of 2 different colours to pairs. Ask: if you have 2 reds and 2 greens what are all the combinations you could get if you took 2 of them, 1 at a time, without looking? This gives a 1-2-1 pattern as with the 2 coins. Point this out to the class then give out **Copymaster 59** and look at it with the children. Get the class back together. Relate the first rows to tossing 1 coin, 2 coins and 3 coins. Now show them how we can generate the triangle by simple addition as shown below.

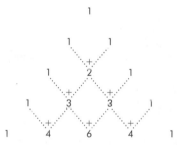

Work with the children on all copymaster challenges.

Closing the lesson 10min

 Use the children's ideas here. Point out, if necessary, that the counting numbers appear, and that the triangle numbers 1, 3, 6, 10 and so on are to be found.

Assessment

Child's performance	Teacher action
Makes some progress but has problems relating outcomes to the pattern in the triangle	Give some more time checking that sequences like the triangle numbers are familiar
Finds a number of patterns but misses some	Spend more time exploring Pascal's Triangle
Does most of the work	Give more investigations

Homework

Ask the children to investigate games involving dice, spinners, or cards (including game cards or playing cards) for probable outcomes: e.g. What are my chances of getting a 'Get out of Jail Free' card in *Monopoly*? Why are the chances of my auntie winning the National Lottery so small?

Investigations

- Make up a track game using what you know about probability and chance. Does it matter where you put reward or punishment squares? Check out games like *Snakes and Ladders* to help you make decisions.

- Working in small groups compose and send letters to manufacturers of items like crisps, biscuits, and sweets. Ask how they make sure the right amount is in each packet. How do they know what size to make the containers for their goods?

- In science, do a temperature experiment. Start with hot water and take the temperature every, say, 3 minutes as it cools. Plot a line graph. What do you notice?

- Set up a database on the computer on a topic you can easily get information about. It could be about, e.g. birds, fish, plants, local buildings, painters and/or paintings, pop stars or even food and drink. How do you decide on your fields? When your friends try out your database can they find all the information they want?

- Carry out a survey on an issue that interests you and your friends. Choose a topic where people might have strong opinions, e.g. on threatened wildlife, whether children watch too much TV, if violent films should be banned, or how children are treated by adults. Design the survey. How will you get people's views? How can you analyse their views and present them using charts and pictorial representation?

- Collect league tables of a favourite sport. Use these to work out averages. If you can get data on individual players then how does their scoring average compare with other players?

- Find out how to write Ordnance Survey map references. With a friend plan a route and write it out just using these references. Can someone else follow your route?

- Collect some data on favourite things (like TV programmes, colours, or snacks) and enter them into a spreadsheet on the computer. Make charts and graphs on screen that can be printed up for you. Remember that discrete data should not be represented as a line graph.

Assessment

- The children should reproduce the data presented below then answer questions of the type indicated.

Number of 'penny' chews	1	2	3	4	5	10
Cost in pence	3	6	9	12	15	30

Can they work out what:
   7 would cost?
   or 9 would cost?
   or 19 would cost?

This is not continuous data but a bar line graph could be used and the relationship found by inspection.

- The children should measure people's hand spans – working out a consistent way of doing this – and then represent their findings as grouped discrete data.

- Make a collection, with the help of the children, of charts and graphs from newspapers and magazines. The children should tell you what they represent and how fair the representation is.

- Using a holiday brochure that gives flight times to far-away destinations get the children to plot a graph like that shown opposite. This is artificial, of course, in that we are assuming a constant speed. However, with atlases, the children can attempt to work out the location of the plane at different times on its journey. As an extension of this get hold of complete national railway timetables. These have times and distances between stops. Get the children to try and plot a graph that includes the stops – so the time will pass but no distance will be covered while the train is stationary. The children can trace their route using an atlas.

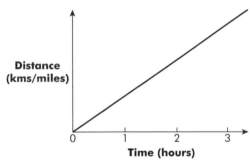

- Get a variety of containers. Get the children to fill each one using the same small beaker, or similar. They should note how many beakers full (or fractionally full if possible) and then represent their findings pictorially.

- Set up a production line of 3 children making birthday cards (see the diagram below). The children should work out how long it takes them to make 3 cards, 5 cards, and 7 cards. Then they should plot a line graph and use it to estimate how long it might take them to make 25 cards.

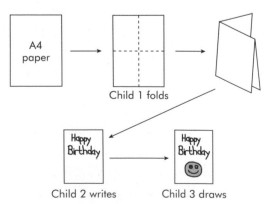

Learning Targets: Number Key Stage 2
RECORD SHEET

Name _____ Class/Year _____ Teacher's initials _____

Section		Theme	Performance in relation to learning targets			Summative remarks
			1	2	3	
1 Identifying numbers	1	Using the number line				
	2	Place value				
	3	Directed numbers				
	4	Number systems, symbols and words				
2 Addition and Subtraction	5	Addition and subtraction to 10				
	6	Addition and subtraction to 20				
	7	Addition of numbers to 100				
	8	Subtraction of numbers to 100				
	9	Addition of numbers greater than 100				
	10	Subtraction of numbers greater than 100				
	11	Decimals				
	12	Addition and subtraction using money				
	13	Number bases				
3 Fractions	14	Halves and quarters				
	15	Developing fractions				
	16	Equivalent fractions				
	17	Fractions, decimals and percentages				
4 Multiplication and Division	18	Multiplication tables				
	19	Multiplication				
	20	Division				
	21	Remainders				
	22	Long multiplication				
	23	Long division				
	24	Multiplying decimals				
	25	Dividing decimals				
	26	Ratio				
	27	Scale				
5 Mental arithmetic and number patterns	28	Addition and subtraction				
	29	Developing strategies				
	30	Missing numbers				
	31	Money				
	32	Factors				
	33	Exploring patterns				
6 Handling data	34	Simple charts				
	35	Frequency and grouped data				
	36	Graphs				
	37	Averages				
	38	Chance				
	39	Co-ordinates				
	40	Fair or unfair				
	41	Probability				

1 Put a counter on 1, then put one on every second number from there until the end of the line. What kind of numbers are the counters on?

_____

2 How could you calculate 3 + 4 using a number line?

Show your method of working here:

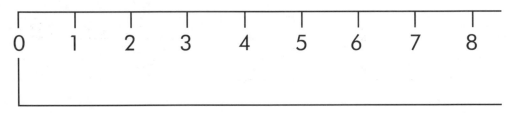

3 Show how to do (4 + 14) − 3 using this number line:

0 1 2 3 4 5 6 7 8 9 10 11 12 13 14 15 16 17 18 19 20

4 Start at 18 and count down in 3s. Write each number you land on here:

18						

5 How could you extend the line to keep going?

_____

6 Now make up some more challenges to try out on a friend

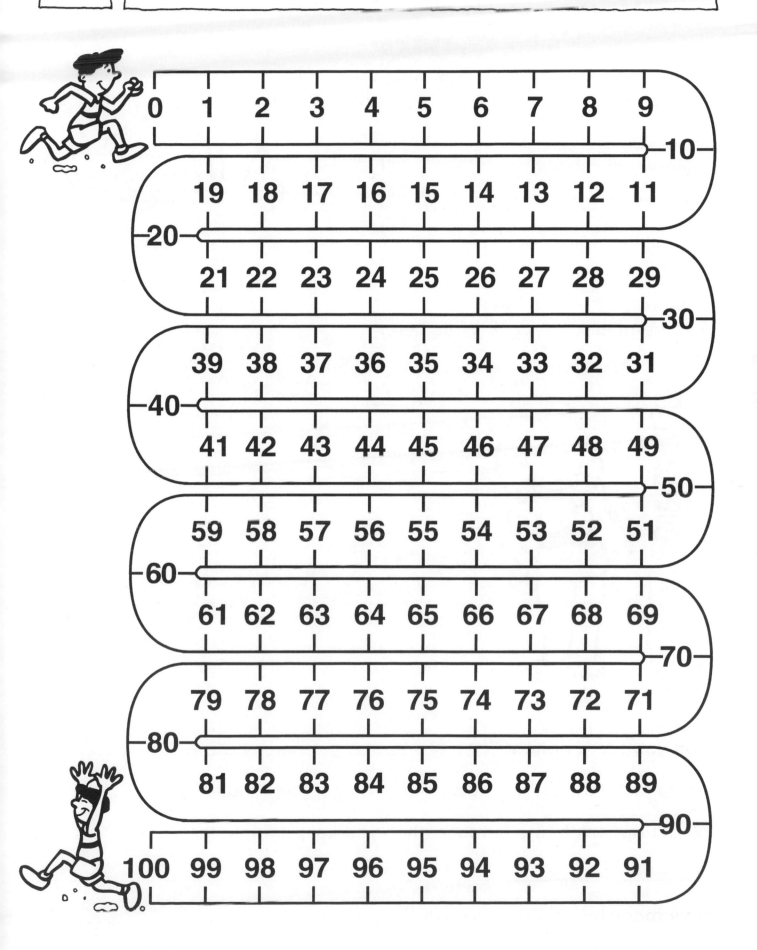

Adding tens

Choose a number between 1 and 9.
Add ten to the number and record your result.
Repeat this process over and over again.

Example: number 7

$$7 + 10 = 17$$
$$17 + 10 = 27$$
$$27 + 10 = 37$$

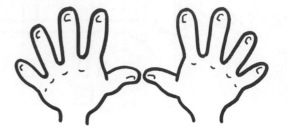

Your number ☐ Write here:

Can I see a pattern?

Can you see a pattern? _____

How many tens do you add before you get three digits? _____

What happens to the number when you keep adding tens?

_____

How many tens make 100, 1000? _____

Temperature

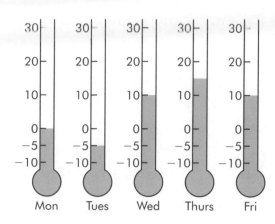

Mon	Tues	Wed	Thurs	Fri	

Which day was the coldest? _____

What was the temperature? _____

Which day was the warmest? _____

What was the temperature? _____

What was the difference between
the warmest and coldest days? _____

On Monday night the temperature was −5°C. During
the day it rose by 15°C. What was the temperature? _____

In the daytime on Tuesday it was 18°C. During the night
the temperature fell by 20°C. What was the temperature? _____

On Wednesday morning it was 5°C but later in the day
the temperature fell by 8°C. What was the temperature? _____

Weather report

Record the temperatures on these days.

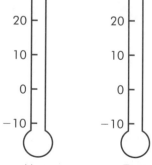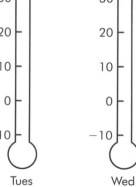

Mon	Tues	Wed	Thurs	Fri

How much has the temperature changed from the first day
to the last? _____

What is the difference between the coldest recorded temperature
and the hottest? _____

Make a graph of your results.

5 | Numbers from around the world

A The Bushmen of Botswana have six words for numbers. They are:

one: *a*	four: *oa-oa*
two: *oa*	five: *oa-oa-a*
three: *ua*	six: *oa-oa-oa*

1 What is the key number in their system? (Remember we have a base 10 system. What do you think theirs is?)

_____

2 Where is Botswana? _____

B The Aztecs used these words:

one: *ce*	six: *chica-ce*
two: *ome*	seven: *chic-ome*
three: *yey*	eight: *chicu-ey*
four: *naui*	nine: *chic-naui*
five: *macuilli*	ten: *maltacti*

1 What do you think is the key number here?_____

2 Why have you chosen that number? _____

3 Where did the Aztecs live? Where in the world is that?

_____

C Here are some words in three different languages, French, German and Latin:

decem	*un*	*ein*	*duo*	*quatre*	*dix*
quinque	*trois*	*tres*	*zehn*	*drei*	*fünf*
cinq	*zwei*	*vier*	*unus*	*deux*	*quattuor*

They are jumbled up, can you draw lines to link them? The words for one, two, three, four, five and ten are written in each of the three languages.

Finger multiplication

Fingers have been used as a calculating aid for hundreds of years. Here are two ways of doing multiplication. Try them out to see if they work and, if they do, think why they work.

A Multiplying by 9

1   Hold up your hands with the palms away from you.

2   To multiply 3 × 9 put your hands like this:

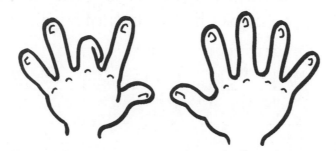

Note that you put the third finger down as it is 3 you want to multiply by.

3   Count the number of upright fingers to the left of the third finger – there are 2. Then count the remaining upright fingers to the right of the 3 (we are calling thumbs fingers here). There are 7, so 3 × 9 = 27.

4   Now multiply some other numbers by 9.

B Multiplying by 6, 7, 8 and 9

Here is an example
  We will calculate 7 × 8.
  Subtract 5 from the 7, leaving 2.
  Raise 2 fingers on your left hand.
  Subtract 5 from 8, leaving 3.
  Raise 3 fingers on your right hand.
  Add the number of raised fingers – in this case 5 (2 + 3).
  Multiply the numbers of raised fingers – in this case 6 (2 × 3).
  So 7 × 8 = 56.

Try some more multiplications using your fingers.

Romans and Egyptians

A Roman numerals

The Roman numerals we want to use are:

1	5	10	50	100	500	1000
I	V	X	L	C	D	M

1   You probably know how to write some other numbers (look at clock faces and reference books if you are not sure). Try writing these:

6	8	9	15	99

2   Doing calculations with Roman numerals is quite hard.

Try working out these sums in our numbers then in Roman numbers:

$$
\begin{array}{r} 63 \\ + 25 \\ \hline \end{array}
\qquad
\begin{array}{r} \text{LXIII} \\ + \text{XXV} \\ \hline \end{array}
$$

Why is it more difficult using Roman numbers? _____

B Egyptian numerals

The symbols we need are:

Egyptian	1	∩	9
Arabic	1	10	100

The Egyptians would have written 236 as   99∩∩∩ $\begin{smallmatrix}111\\111\end{smallmatrix}$

1   Try writing your birth date in Egyptian numerals.
2   Now try this sum using Egyptian numerals: 153 + 162.
3   Make up some calculations of your own. Can you do a multiplication?

A There are special words used to talk about different sizes of groups of musicians. Fill in the missing spaces here:

Groups of musicians

1	Soloist
2	
3	
4	
	Quintet
6	
	Septet
8	

B The following table lists prefixes (beginnings) for words which describe numbers of things.

Complete the table with examples of words which use these prefixes

Prefix	Number	Examples
Bi		
Tri		
Quad		
Deci		
Penta		
Uni		
Hexa		

The winner is the first to put counters on 4 numbers in a row.

1	2	3	4	5	6
7	8	9	10	1	2
3	4	5	6	7	8
9	10	1	2	3	4
5	6	7	8	9	10
1	2	3	4	5	6

Who's the spinner winner?

Paint spots (1)

Spots of paint have splashed onto this sheet of calculations. Can you put in the missing numbers and signs which are covered by paint?

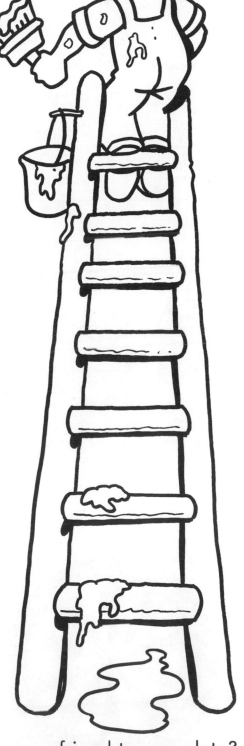

$3 + 3 = 6$

$4 + 1 = 5$

$4 - 2 = $

$10 - = 8$

$9 1 = 10$

$ + 5 = 6$

$ + 2 = 9$

$8 3 = 5$

$2 + = 8$

$0 + 5 = $

$10 - = 0$

$5 2 = 7$

$6 - = 4$

$1 6 = 7$

$ - 1 = 1$

$ + 5 = 8$

$5 3 = 2$

$6 - = 3$

$7 + 3 = $

$8 - 5 = $

$9 - = 9$

$3 6 = 9$

$5 5 = 0$

$10 6 = 4$

Can you make up some paint spot problems for your friend to complete?

Dartboard

106

Paint spots (2)

Spots of paint have splashed onto this sheet of calculations. Can you put in the missing numbers which are covered by paint?

$26 + 14 = $ $ + 25 = 52$ $ + 28 = 61$

$25 + 27 = $ $60 + = 80$ $ + 12 = 41$

$34 + = 50$ $49 + 21 = $ $ + 33 = 84$

$42 + = 59$ $ + 18 = 54$ $24 + 37 = $

$16 + 16 = $ $ + 36 = 73$ $13 + 18 = $

$63 + 21 = $ $16 + = 42$ $ + 35 = 71$

$17 + = 35$ $ + 79 = 91$ $88 + = 99$

Make up some paint spot problems for your friend to complete. Check that you both agree the answers!

Turning numbers around

Here are some calculations made with these two numbers:

$$17$$

$$25$$

$$17 + 8 = 25 \qquad 25 - 17 = 8 \qquad 25 - 8 = 17$$

Look at these examples made using 35 and 23:

$$35$$

$$23$$

$$35 + 23 = 58 \qquad 58 - 35 = 23 \qquad 58 - 23 = 35$$

Record your problems here.

Is this the problem page then?

14 Estimations before calculations

Complete these calculations by first rounding the numbers to ten then to hundreds.
Here is an example.

459 + 397
Rounding to nearest 10
460 + 400 = 860
Rounding to nearest 100
500 + 400 = 900
Actual answer
459 + 397 = 856
Which is the better estimate?

231 + 589
Rounding to nearest 10
____ + ____ = ____
Rounding to nearest 100
____ + ____ = ____
Actual answer
231 + 589 =
Which is the better estimate?

346 + 237
Rounding to nearest 10
350 + 240 = ____
Rounding to nearest 100
300 + ____ = ____
Actual answer
346 + 237 =
Which is the better estimate?

671 + 198
Rounding to nearest 10
____ + ____ = ____
Rounding to nearest 100
____ + ____ = ____
Actual answer
671 + 198 =
Which is the better estimate?

349 + 123
Rounding to nearest 10
____ + ____ = ____
Rounding to nearest 100
____ + ____ = ____
Actual answer
349 + 123 =
Which is the better estimate?

357 + 286
Rounding to nearest 10
____ + ____ = ____
Rounding to nearest 100
____ + ____ = ____
Actual answer
357 + 286 =
Which is the better estimate?

Flats, longs and units

263 − 185

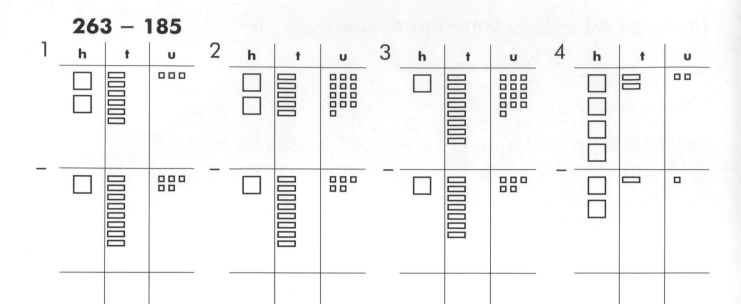

1 | h | t | u |
2 | h | t | u |
3 | h | t | u |
4 | h | t | u |

5 | h | t | u |
6 | h | t | u |
7 | h | t | u |
8 | h | t | u |

9				10				11				12		
h	**t**	**u**		**h**	**t**	**u**		**h**	**t**	**u**		**h**	**t**	**u**
5	3	7		4	7	2		3	8	4		8	1	1
− 2	4	8		− 1	9	6		− 2	9	6		− 1	4	3

There are 80 ways to solve this puzzle!

How many solutions can you find?

In each example you must use all the digits 0–9 once only.

You must not put the zero in the thousands column.

The first one is done for you.

```
  1   0   8   9
-     7   6   5
  ─────────────
      3   2   4
```

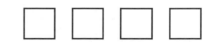

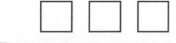

17 Decimal sums

This one is done for you.

units	tenths
▭	▫▫▫

+

units	tenths
▭	▫▫▫▫

units	tenths
▭▭	▫▫▫▫▫ ▫▫

units	tenths
▭	▫▫

+

units	tenths
▭	▫▫▫▫

units	tenths
▭▭	▫▫▫▫

+

units	tenths
▭	▫▫▫▫

units	tenths
▭	▫▫▫▫▫ ▫

+

units	tenths
▭	▫▫▫▫▫

units	tenths
▭	▫▫▫▫▫

+

units	tenths
▭	▫▫▫▫▫ ▫▫

units	tenths
▭	▫▫▫▫

+

units	tenths
▭	▫▫▫▫▫ ▫

Complete the following sums using Multi-base® to help you.

units	tenths
2	5
+1	2

units	tenths
1	7
+2	2

units	tenths
3	3
+1	6

units	tenths
5	3
+1	9

units	tenths
1	8
+2	5

units	tenths
0	9
+1	2

units	tenths
1	7
+2	3

units	tenths
1	8
+4	5

units	tenths
2	2
+1	9

units	tenths
4	7
+3	4

Now try these sums.

units . tenths

3 . 4
+2 . 6
___ . ___

units . tenths

3 . 7
+4 . 2
___ . ___

units . tenths

4 . 5
+3 . 2
___ . ___

units . tenths hundredths

6 . 2   4
+2 . 4   5
___ . ___ ___

£1.00 square

Start

£0.01	£0.02	£0.03	£0.04	£0.05	£0.06	£0.07	£0.08	£0.09	£0.10
£0.11	£0.12	£0.13	£0.14	£0.15	£0.16	£0.17	£0.18	£0.19	£0.20
£0.21	£0.22	£0.23	£0.24	£0.25	£0.26	£0.27	£0.28	£0.29	£0.30
£0.31	£0.32	£0.33	£0.34	£0.35	£0.36	£0.37	£0.38	£0.39	£0.40
£0.41	£0.42	£0.43	£0.44	£0.45	£0.46	£0.47	£0.48	£0.49	£0.50
£0.51	£0.52	£0.53	£0.54	£0.55	£0.56	£0.57	£0.58	£0.59	£0.60
£0.61	£0.62	£0.63	£0.64	£0.65	£0.66	£0.67	£0.68	£0.69	£0.70
£0.71	£0.72	£0.73	£0.74	£0.75	£0.76	£0.77	£0.78	£0.79	£0.80
£0.81	£0.82	£0.83	£0.84	£0.85	£0.86	£0.87	£0.88	£0.89	£0.90
£0.91	£0.92	£0.93	£0.94	£0.95	£0.96	£0.97	£0.98	£0.99	£1.00

Remember to use base eight here.

$1 + 7 = \boxed{}$

$4 + 4 = \boxed{}$

$5 + 12 = \boxed{}$

$3 + \boxed{} = 10$

$10 + 15 = \boxed{}$

$17 + 12 = \boxed{}$

$\boxed{} + 12 =$

$\boxed{} + 21 = 30$

$3 + 12 = \boxed{}$

$7 + 7 = \boxed{}$

$2 + 17 = \boxed{}$

$\boxed{} + 6 = 12$

$6 + 11 = \boxed{}$

$11 + 11 = \boxed{}$

$2 + \boxed{} = 21$

$17 + \boxed{} = 25$

How many ways can you make 15 using spider addition and subtraction?

$7_{\text{(base 8)}} + 6_{\text{(base 8)}} = 15_{\text{(base 8)}}$

$27_{\text{(base 8)}} - 12_{\text{(base 8)}} = 15_{\text{(base 8)}}$

$30_{\text{(base 8)}} - 13_{\text{(base 8)}} = 15_{\text{(base 8)}}$

12 doubled is 24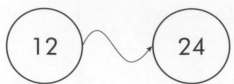

Draw an arrow to connect each number with its 'double'.

Double trouble

10 8
2 4 12
18 6 20
16 24

   5 is half of 10

Connect the numbers below using 'halving'.

2 22
30 6
10 8 40
18 20
12

5
20 10
3 9 1
11 4
15 6

Write in halved or doubled .

Half the fun   16 ⬜ 8   36 ⬜ 18

Divide these shapes into halves.

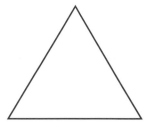

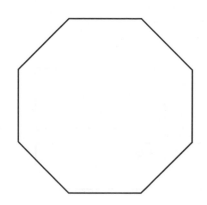
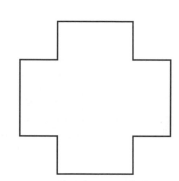
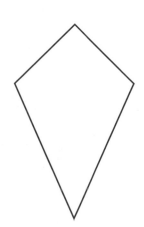

Now see which ones you can divide into quarters.

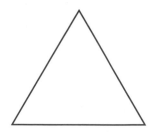

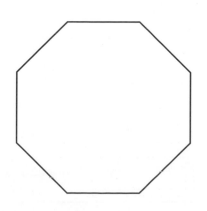
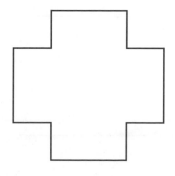
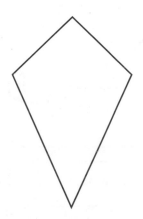

Fractions in colour

Colour a whole, $\frac{1}{2}$, $\frac{1}{4}$, $\frac{1}{8}$, $\frac{1}{10}$, $\frac{1}{5}$.

Colour $\frac{3}{10}$, $\frac{2}{5}$, $\frac{5}{8}$.

Can you make more different fractions of a whole?

Counter challenges

A Take 6 counters. How many different ways can you group them? There are 2 here, find some others:

> 1 group of 6
> 6 groups of 1

> Now write some fractions of sixths. For example $\frac{1}{6}$ and $\frac{5}{6}$.

$\frac{3}{6}$ is the same as $\frac{1}{2}$ – use your counters to see why. Are there more of your 'sixth' fractions that can be written as other fractions?

> Write them down here.

B Try these sorts of challenges again using:

- 10 counters
- 8 counters
- 5 counters.

What do you notice?
Write some of your ideas here:

Crikey – I've lost count!

Make your mark

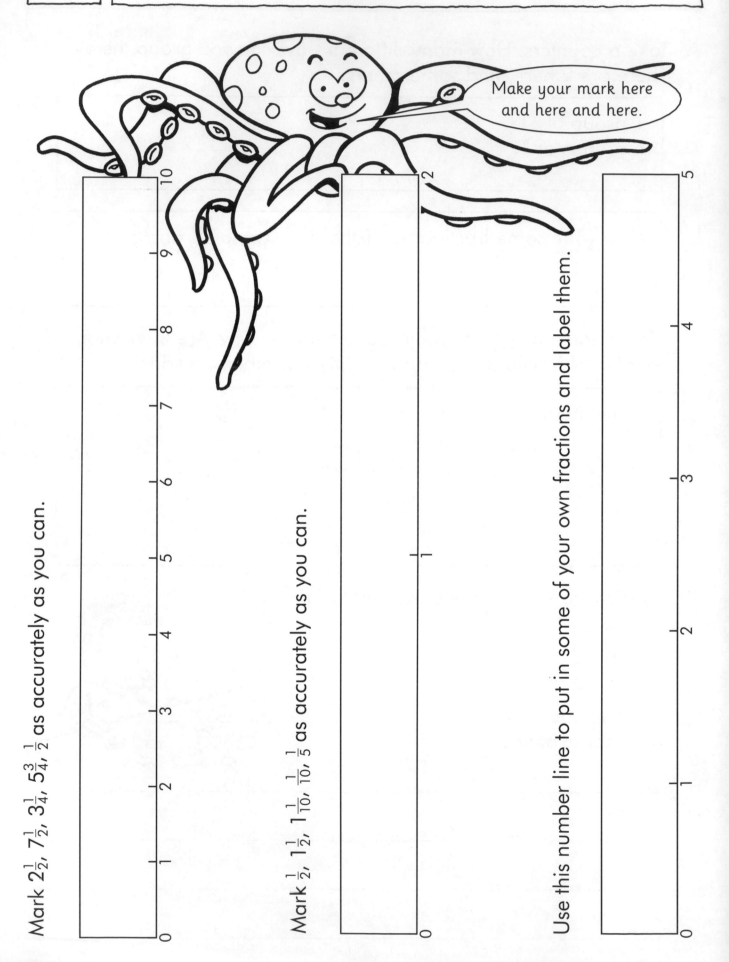

Make your mark here and here and here.

Mark $2\frac{1}{2}$, $7\frac{1}{2}$, $3\frac{1}{4}$, $5\frac{3}{4}$, $\frac{1}{2}$ as accurately as you can.

Mark $\frac{1}{2}$, $1\frac{1}{2}$, $1\frac{1}{10}$, $\frac{1}{10}$, $\frac{1}{5}$ as accurately as you can.

Use this number line to put in some of your own fractions and label them.

$\dfrac{1}{2}$	$\dfrac{1}{3}$	$\dfrac{1}{4}$
$\dfrac{1}{5}$	$\dfrac{1}{6}$	$\dfrac{1}{8}$
$\dfrac{1}{10}$	$\dfrac{1}{12}$	$\dfrac{3}{4}$
$\dfrac{2}{3}$	$\dfrac{2}{5}$	$\dfrac{3}{5}$
$\dfrac{5}{6}$	$\dfrac{3}{8}$	$\dfrac{5}{8}$
$\dfrac{7}{8}$	$\dfrac{3}{10}$	$\dfrac{7}{10}$
$\dfrac{9}{10}$	$\dfrac{5}{10}$	$\dfrac{7}{12}$
$\dfrac{11}{12}$		

What fraction of me is tail, I wonder?

What's missing?

Fill in the missing fractions and percentages.

0		$\frac{1}{5}$	$\frac{1}{2}$		
0%	10%			75%	100%

Now try these – they are decimal fractions and percentages.

Be careful – we are certainly not armless!

0		0.4	0.5	0.75		1.0
0	10%				80%	

This table shows decimals, percentages and fractions. What's missing?

$\frac{1}{2}$		$\frac{1}{4}$		$\frac{2}{5}$
0.5	0.75			
50%			20%	

Use decimals, percentages and fractions to make a table of your own.

Multiplication triangle

A

$6 \times 7 =$ [　　]

$9 \times 6 =$ [　　]

$7 \times 8 =$ [　　]

$2 \times 3 \times 2 =$ [　　]

$3 \times 5 \times 2 =$ [　　]

$10 \times 6 \times 3 =$ [　　]

B Take care with the brackets in some of these.

A bookshelf bracket, will this do I wonder?

$12 \times 2 \times 3 =$ [　　]

$2 \times 2 \times 2 \times 2 =$ [　　]

$6 \times 8 \times 7 =$ [　　]

$2 \times (3 \times 4) =$ [　　]

$(4 \times 9) + 3 =$ [　　]

$(1 \times 11) - 5 =$ [　　]

C Put in the brackets and then do the calculation.

$2 \times 3 + 7 =$ [　　]

$4 \times 6 \times 12 =$ [　　]

$4 \times 10 + 5 =$ [　　]

$5 \times 3 - 7 =$ [　　]

$2 + 12 \times 3 =$ [　　]

$10 \times 3 + 6 \times 7 =$ [　　]

Solve these problems.

$6 \times 7 =$ ☐ $3 \times 2 \times 2 =$ ☐

$4 \times 12 =$ ☐ $13 \times 3 \times 3 =$ ☐

How did you do them? Write down some methods.

I think this solution is pretty good!

Find the solutions to these.

$10 \div 2 =$ ☐ $6 \times 4 =$ ☐

$2 \times$ ☐ $= 10$ ☐ $\div 4 = 6$

$8 \times 3 =$ ☐   ☐ $\div 8 = 3$

How are multiplication and division connected?

Make up some more of these calculations.
Write them as multiplications and then divisions.

☐ $\times$ ☐ $=$ ☐   ☐ $\times$ ☐ $=$ ☐   ☐ $\times$ ☐ $=$ ☐

☐ $\div$ ☐ $=$ ☐   ☐ $\div$ ☐ $=$ ☐   ☐ $\div$ ☐ $=$ ☐

Number cards

Glue onto thin card and cut out.

The top row are rounding cards – 1 per group. The others should be shared among the groups. ✂

To nearest 10	To nearest 100	To nearest 100	To nearest 1000
48	32	17	96
173	611	41	102
7622	33	119	2081
3704	136	727	198
148	223	222	5909
744	4388	151	379
58	439	1005	552
149	53	37	7112
182	582	169	7802
6743	9999	19	979
28	802	721	44
92	561	88	111

Here are two methods of doing a long multiplication.

52 × 13

Method 1

First, multiply 3 by 52.	Then multiply 10 by 52.	Then add the two products together
52 × **13** 156	**52** × **13** 156 520	52 × 13 **156** **520** 676

Method 2

52 × 13 is the same as
50 × 13 plus **2** × 13
50 × **13** is the same as
50 × **10** plus 50 × **3**

$$50 \times 10 = 500$$
$$50 \times \ 3 = 150$$
$$\overline{650}$$ Add the products together

and $2 \times 13 = \underline{\ 26}$
$$676$$ Add the products together

I've got an answer.

I've got an answer too!

Try these two methods on these problems:

64 × 27   15 × 32   18 × 89   31 × 19

Now try to solve these multiplications using the method you prefer.

123 × 45   189 × 22   352 × 16   406 × 13

Multiplication methods

25 × 13

$$25 \times \textcircled{1} = 25$$
$$25 \times 2 = 50$$
$$25 \times \textcircled{4} = 100$$
$$25 \times \textcircled{8} = 200$$

Total to 13

What is happening to these numbers each time?

So
$$25 \times 1 = 25$$
$$25 \times 4 = 100$$
$$25 \times 8 = \underline{200}$$
$$\underline{325}$$

So 25 × 13 = 325

Try this method for these:

15 × 22   34 × 9   19 × 12   26 × 26

Now do them using another method you know.

How do the two methods compare?

Zoo food

Grain supplies

The lorry carries 510 sacks. 17 sacks are needed each day.

How many days' supply on a lorry? _____

The store contains 714 sacks already.

How many days' supply is that?_____

Fish

Fish comes by the kilo.
884 kilos came yesterday.
How many kilos could each of the 68 penguins have? _____

Fruit and vegetables

There are 23 orangutans. They like oranges and have at least one each
a day. There are 345 oranges in store.
In how many days will the keeper need new supplies? _____

Meat

The meat-eating animals have about a kilogram each (on average). The
zoo needs 89 kilograms a day. How many days' supply here?

8010 kg _____   115 kg _____   5518 kg _____

Long multiplication (2)

Use long multiplication to complete all these.

Then use a calculator to check the answers.

$155 \times 2.5 = \boxed{}$

$276 \times 5.7 = \boxed{}$

Wow! This is a long multiplication!

$76 \times 3.72 = \boxed{}$

$801 \times 4.2 = \boxed{}$

$651 \times 4.8 = \boxed{}$

$5.23 \times 63 = \boxed{}$

$5.9 \times 301 = \boxed{}$

$9.5 \times 167 = \boxed{}$

$8.7 \times 528 = \boxed{}$

$10.25 \times 18 = \boxed{}$

Do you get the point?

Try these divisions.

$18 \div 5 =$ ▢

$78 \div 12 =$ ▢

$9 \div 45 =$ ▢

$54 \div 45 =$ ▢

$22 \div 20 =$ ▢

Now use a calculator to find other examples that have an answer with one decimal point.

▢ $\div$ ▢ $=$ ▢

▢ $\div$ ▢ $=$ ▢

▢ $\div$ ▢ $=$ ▢   ▢ $\div$ ▢ $=$ ▢

▢ $\div$ ▢ $=$ ▢   ▢ $\div$ ▢ $=$ ▢

Dividing decimals

1   6)15.6

2   3)43.8

3   7)25.9

4   9)391.5

5   13)390

6   43)1161

7   20)246.4

8   30)342.6

9   50)146.5

10 18)205.2

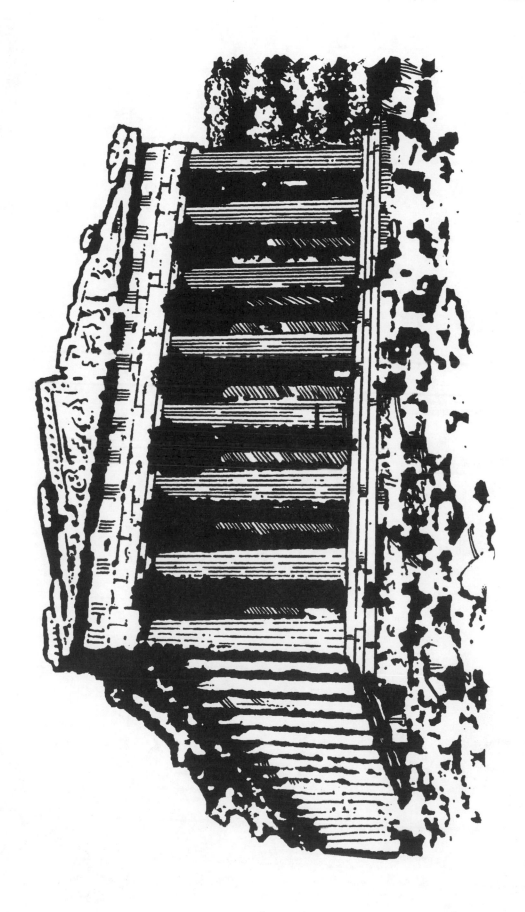

Addition squares

Complete these addition squares.

+	4	5	6
4			
5			
6			

+	1	3	5
2			
4			
6			

Are there any patterns? _____

Do you notice anything else?

_____

Look for the patterns when you have completed this square.

+	1	2	3	4	5	6
1						
2						
3						
4						
5						
6						

Now make up some of your own.

+			

+			

Try these additions.

A 12 + 9   D 5 + 7   G 7 + 18

B 15 + 8   E 14 + 7   H 16 + 9

C 13 + 8   F 19 + 3   I 15 + 9

Write out how you did them. For example for the first one you might have done 10 + 9 then add 2.

	My way	Can you write in another way?
A		
B		
C		
D		
E		
F		
G		
H		
I		

Cards for Bingo

6	40	4	5	2
24	25	11	3	27
17	19	16	21	49
12	3	36	29	81
56	1	99	13	43
16	48	63	16	7

Bingo!

2×12		$(8 + 5) - 2$	$13 - 7$		$9 + 7$
$3 + 4 + 5$		$8 + 8$		7×8	
	$12 - 9$		3×9		8×2

Reveal your maths experience. Is your number up?

Is this a cover up?

	$6 + 3 + 10$	$12 - 9$		$(9 + 3) - 8$	6×8
9×7	5×8		5×5	7×3	
		4×9			9×9

Function machines

Draw 4 of your own machines here and make them work.

Outdoor world
Cost of sleeping and cooking gear for one camper.

Food heaven
Cost of cake, crisps and squash for a visiting class from another school.

Party bag
Fill 15 party bags with a blower, balloon, small toy and chocolate bar.

Model mania
Cost of competition prizes for a model maker – limit £75.

Back to school
School bag, stationery and pen and pencil needed for new term. How much?

Sleepover
Pack an overnight bag – pyjamas, washbag, comic and cuddly toy. How much?

Large 100 square

1	2	3	4	5	6	7	8	9	10
11	12	13	14	15	16	17	18	19	20
21	22	23	24	25	26	27	28	29	30
31	32	33	34	35	36	37	38	39	40
41	42	43	44	45	46	47	48	49	50
51	52	53	54	55	56	57	58	59	60
61	62	63	64	65	66	67	68	69	70
71	72	73	74	75	76	77	78	79	80
81	82	83	84	85	86	87	88	89	90
91	92	93	94	95	96	97	98	99	100

Magic squares

It must
be magic!

It mouse-t
be magic!

In a class of 38 children the birthdays
fall in these months:

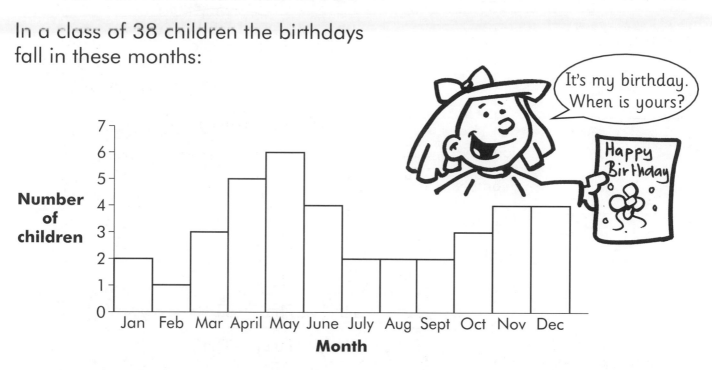

Check that there are 38 children in the class then answer these
questions:

1 Which month has most birthdays? _____

2 Which month has least birthdays? _____

3 How many months have 4 children in each? _____

4 What are these months? _____

5 Which months have 2 children's birthdays in each? _____

_____

Now make up some questions of your own.

_____

_____

_____

_____

_____

Sailings from the mainland to the Isle of Sun

6 sailings   3 sailings

January	
February	
March	
April	
May	
June	
July	
August	
September	
October	
November	
December	

Frequency of wearing a cardigan or sweater at 10 am on each of 5 days

Sample: 30 children

2 people

Monday	
Tuesday	
Wednesday	
Thursday	
Friday	

Colour the pie charts using a different
colour for each section:
What children do for lunch

_____

_____

_____

_____

Packed
lunch

School
meals

Go home

Land use in a town

_____

_____

_____

_____

Parks

Homes

Woods

Shops

Business

Ingredients in a soft drink

_____

_____

_____

_____

Carbon
dioxide

Flavourings

Sugar

Water

What is the median average of these?
(Remember to put them in order first).

A   22   27   19   23   24   ☐

B   8   9   5   1   3   ☐

C   3   5   13   12   11   9   4   ☐

D   1   1   5   5   3   ☐

E   1   12   5   8   6   13   ☐

F   5   9   15   13   10   11   8   8   ☐

What is the mode average of these?
(Do not forget to look for the most common.)

G   1   1   3   5   7   1   7   ☐

H   23   22   21   20   20   23   23   21   ☐

I   15   18   18   17   17   17   18   19   18   ☐

J   1   100   100   1   10   10   10   100   10   ☐

How did you get on?
Can you do them mentally?
Wow! What brain exercise!

Mean, median and mode

Calculate the mean average of these:

A 4,   7,   5,   5,   9   ☐

B 15,   18,   12,   10,   15,   14   ☐

C 1,   1,   3,   7,   3,   4,   2   ☐

D 105,   52,   122,   378,   190,   201,   149   ☐

Find the mode and mean of these:

2, 7, 3, 4, 9, 8, 9, 6, 9, 4, 2, 9

mode ☐   mean ☐

Find the median and mean of these salaries:

 £79,000   £20,000   £11,000   £10,000   £5,000

median ☐   mean ☐

Use a calculator if you need to where you see one shown.

What are the ranges here?

A   1, 3, 19, 17, 2, 5, 9

B   1, 1, 5, 5, 3

C   17, 201, 15, 57, 84, 198, 206

D   1, 1, 1, 5, 5, 5

Now try this using a calculator.

Harry's Haulage is a removal company based in Birmingham. Some of the removals the company has done recently are (in miles):
London 105   Bristol 65   Edinburgh 270
Nottingham 70   Liverpool 100

What is the range?

What is the mean average distance travelled?

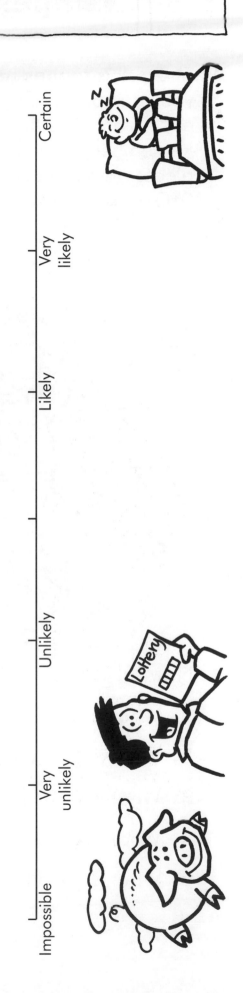

A The task is to produce a temperature conversion graph with which you can convert degrees Celsius (°C) and degrees Fahrenheit (°F) to each other. Use the graph paper your teacher has given you. Remember to think about the **scale** for each axis. Here are two pieces of information you need:
0 °C is equivalent to 32 °F and 100 °C is equivalent to 212 °F.

B Now solve these using your graph.

1   What is 10 °C in °F? _____

2   What is 100 °F in °C? _____

3   What is 20 °C in °F? _____

4   What is the difference between 45 °F and 85 °F in °C? _____

5   What is the difference between 15 °C and 30 °C in °F? _____

Game track

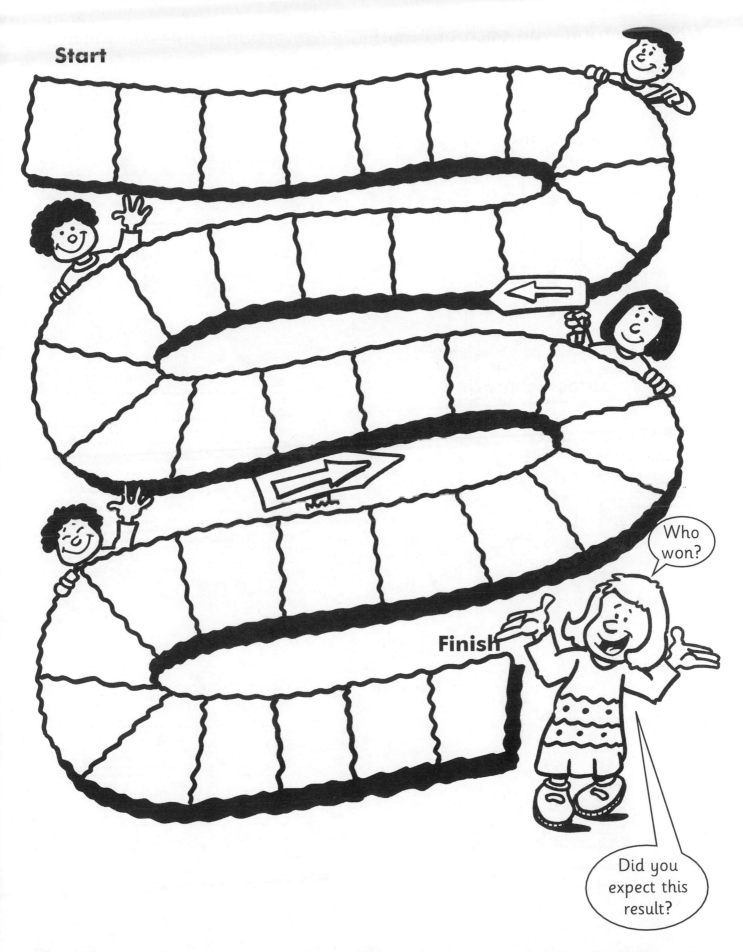

Spot the errors

What is wrong with each of these?

Holidays

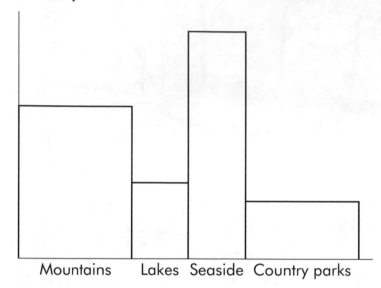

1 _____

_____

_____

_____

Favourite cartoon characters

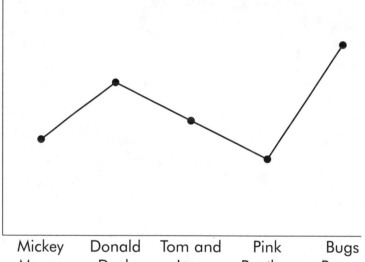

2 _____

_____

_____

_____

Numbers of felt-tip pens

3 _____

_____

_____

_____

Put these on the probability line and add some of your own.

1

0.5

0

Can I have some seconds

It will snow on my home in July.	My favourite football team will win the League this year.	The Prime Minister will visit our school next year.
One of our teachers will leave next year.	There will be two weeks of hot, sunny weather in June next year.	I can pick the Number 1 single for next week.

Can you find patterns in the rows of numbers that you have found when tossing coins?

Is the pattern of outcomes when 4 coins are tossed shown here?

What does each row of numbers add up to? Is there a pattern?

What other patterns can you see or make from this array?

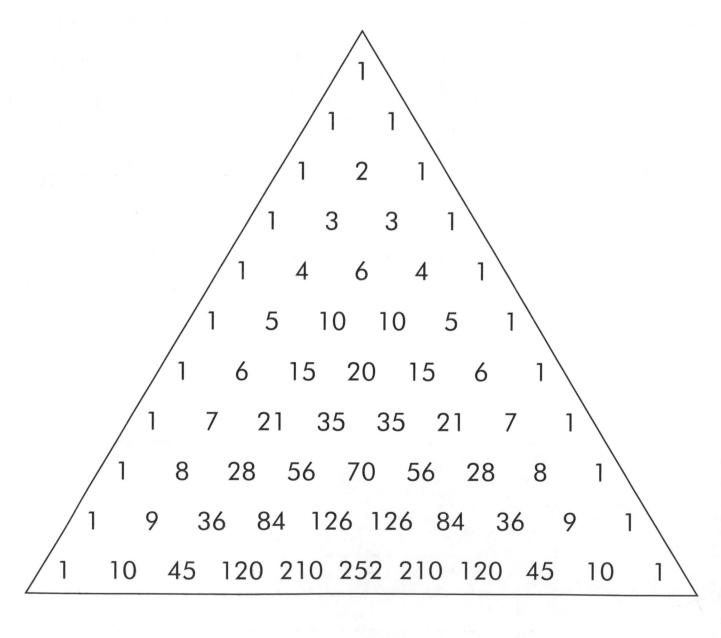

```
                    1
                 1     1
              1     2     1
           1     3     3     1
        1     4     6     4     1
     1     5    10    10     5     1
  1     6    15    20    15     6     1
1     7    21    35    35    21     7     1
1   8   28   56   70   56   28   8   1
1   9   36   84  126  126   84   36   9   1
1  10   45  120  210  252  210  120  45  10   1
```

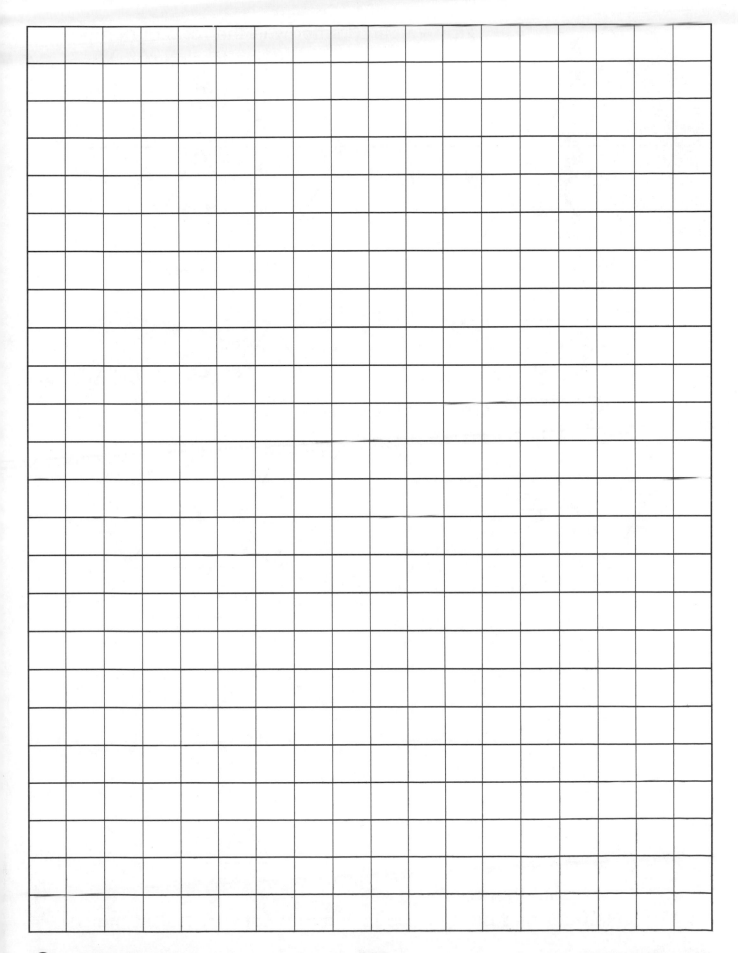

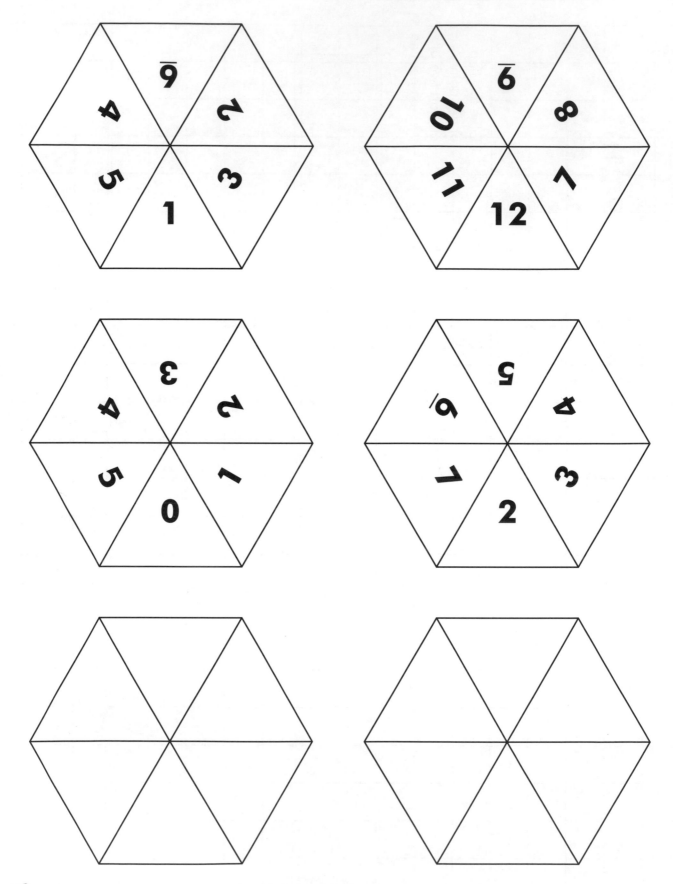

 Stick onto thin card. Cut out.
Number and pierce through centre point with a used match.

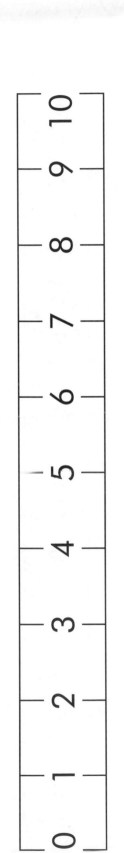

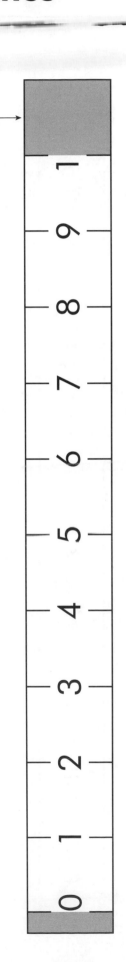

Glue this flap under point A on part two of the number line →

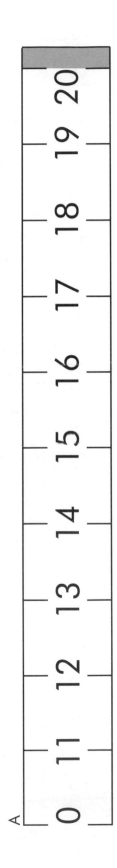

Number line to 100

0 1 2 3 4 5 6 7 8 9 10 11 12 13 14 15 16 17 18 19 2

0 21 22 23 24 25 26 27 28 29 30 31 32 33 34 35 36 37 38 39 4

0 41 42 43 44 45 46 47 48 49 50 51 52 53 54 55 56 57 58 59 6

0 61 62 63 64 65 66 67 68 69 70 71 72 73 74 75 76 77 78 79 8

0 81 82 83 84 85 86 87 88 89 90 91 92 93 94 95 96 97 98 99 100

0	1	2	3	4	5
6	7	8	9	10	11
12	13	14	15	16	17
18	19	20	21	22	23
24	25	26	27	28	29
30	31	32	33	34	35
36	37	38	39	40	41
42	43	44	45	46	47
48	49	50			

51	52	53	54	55	56
57	58	59	60	61	62
63	64	65	66	67	68
69	70	71	72	73	74
75	76	77	78	79	80
81	82	83	84	85	86
87	88	89	90	91	92
93	94	95	96	97	98
99	100				

0	**1**
2	**3**
4	**5**
6	**7**
8	**9**

Blank 100 squares

Multiplication squares

✕

✕

✕

✕

✕

✕

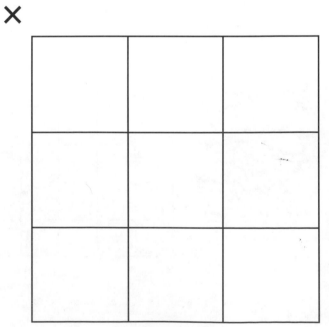

Triangle dotty paper

Graph paper (2 mm/10 mm)

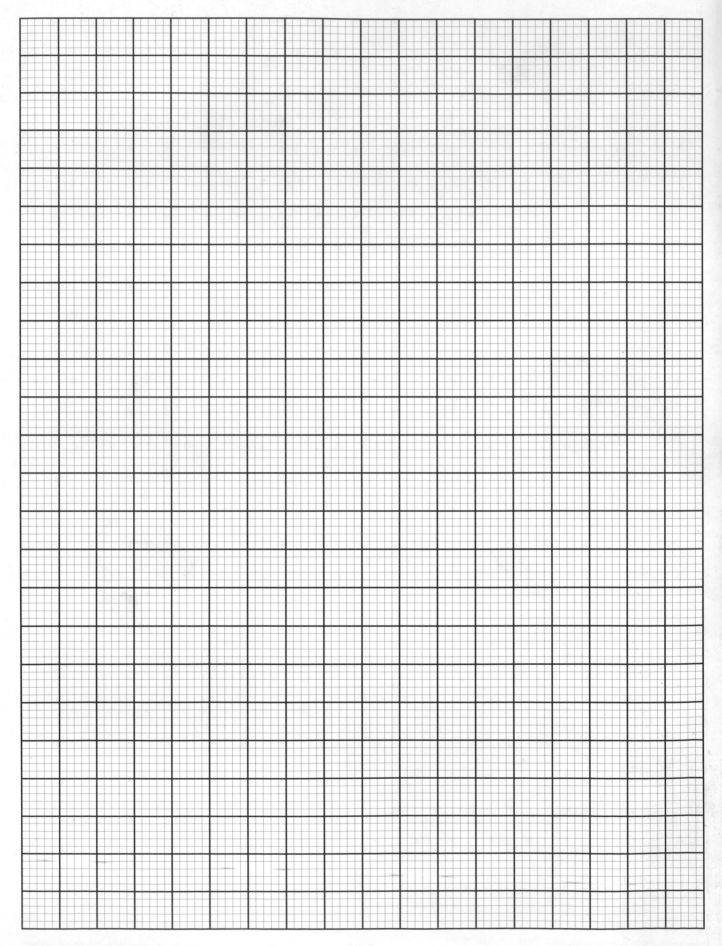

166